Koichi Asano
Mass Transfer

Related Titles

Gmehling, J., Menke, J., Krafczyk, J., Fischer, K.

Azeotropic Data

2nd Ed.

2004
ISBN 3-527-30833-4

Benitez, J.

Principles and Modern Applications of Mass Transfer Operations

2002
ISBN 0-471-20344-0

Sundmacher, K., Kienle, A. (eds.)

Reactive Distillation

Status and Future Directions

2003
ISBN 3-527-30579-3

Sanchez Marcano, J. G., Tsotsis, T. T.

Catalytic Membranes and Membrane Reactors

2002
ISBN 3-527-30277-8

Incropera, F. P., DeWitt, D. P.

Fundamentals of Heat and Mass Transfer

2002
ISBN 0-471-38650-2

Pereira Nunes, S., Peinemann, K.-V. (eds.)

Membrane Technology

in the Chemical Industry

2001
ISBN 3-527-28485-0

Ingham, J., Dunn, I. J., Heinzle, E., Prenosil, J. E.

Chemical Engineering Dynamics

An Introduction to Modelling and Computer Simulatio

2000
ISBN 3-527-29776-6

Incropera, FPI

Fundamentals of Heat & Mass Transfer, 4. ed. + Interactive Heat Transfer Set

1996
ISBN 0-471-15374-5

Rushton, A., Ward, A. S., Holdich, R. G.

Solid-Liquid Filtration and Separation Technology

1993
ISBN 3-527-28613-6

Koichi Asano

Mass Transfer

From Fundamentals to Modern Industrial Applications

WILEY-VCH Verlag GmbH & Co. KGaA

The Author

Koichi Asano
Tokyo Institute of Technology
121-1 Ookayama 2-chome
Meguro-ku, Tokyo
Japan

■ All books published by Wiley-VCH are carefully produced. Nevertheless, authors, editors, and publisher do not warrant the information contained in these books, including this book, to be free of errors. Readers are advised to keep in mind that statements, data, illustrations, procedural details or other items may inadvertently be inaccurate.

Library of Congress Card No.:
applied for

British Library Cataloguing-in-Publication Data
A catalogue record for this book is available from the British Library

Bibliographic information published by Die Deutsche Bibliothek
Die Deutsche Bibliothek lists this publication in the Deutsche Nationalbibliografie; detailed bibliographic data is available in the Internet at <http://dnb.ddb.de>

© 2006 WILEY-VCH Verlag GmbH & Co. KGaA, Weinheim, Germany

All rights reserved (including those of translation into other languages). No part of this book may be reproduced in any form – by photoprinting, microfilm, or any other means – nor transmitted or translated into a machine language without written permission from the publishers. Registered names, trademarks, etc. used in this book, even when not specifically marked as such, are not to be considered unprotected by law.

Printed in the Federal Republic of Germany
Printed on acid-free paper

Composition ProSatz Unger, Weinheim
Printing betz-druck GmbH, Darmstadt
Bookbinding Litges & Dopf Buchbinderei GmbH, Heppenheim

ISBN-13: 978-3-527-31460-7
ISBN-10: 3-527-31460-1

Contents

Preface *XIII*

1 **Introduction** *1*
1.1 The Beginnings of Mass Transfer *1*
1.2 Characteristics of Mass Transfer *2*
1.3 Three Fundamental Laws of Transport Phenomena *3*
1.3.1 Newton's Law of Viscosity *3*
1.3.2 Fourier's Law of Heat Conduction *4*
1.3.3 Fick's Law of Diffusion *5*
1.4 Summary of Phase Equilibria in Gas-Liquid Systems *6*
References *7*

2 **Diffusion and Mass Transfer** *9*
2.1 Motion of Molecules and Diffusion *9*
2.1.1 Diffusion Phenomena *9*
2.1.2 Definition of Diffusional Flux and Reference Velocity of Diffusion *10*
2.1.3 Binary Diffusion Flux *12*
2.2 Diffusion Coefficients *14*
2.2.1 Binary Diffusion Coefficients in the Gas Phase *14*
2.2.2 Multicomponent Diffusion Coefficients in the Gas Phase *15*
Example 2.1 *16*
Solution *16*
2.3 Rates of Mass Transfer *16*
2.3.1 Definition of Mass Flux *16*
2.3.2 Unidirectional Diffusion in Binary Mass Transfer *18*
2.3.3 Equimolal Counterdiffusion *18*
2.3.4 Convective Mass Flux for Mass Transfer in a Mixture of Vapors *20*
Example 2.2 *21*
Solution *21*
2.4 Mass Transfer Coefficients *21*
Example 2.3 *24*
Solution *24*

2.5	Overall Mass Transfer Coefficients 24

References 26

3	**Governing Equations of Mass Transfer** 27
3.1	Laminar and Turbulent Flow 27
3.2	Continuity Equation and Diffusion Equation 28
3.2.1	Continuity Equation 28
3.2.2	Diffusion Equation in Terms of Mass Fraction 29
3.2.3	Diffusion Equation in Terms of Mole Fraction 31
3.3	Equation of Motion and Energy Equation 33
3.3.1	The Equation of Motion (Navier–Stokes Equation) 33
3.3.2	The Energy Equation 33
3.3.3	Governing Equations in Cylindrical and Spherical Coordinates 33
3.4	Some Approximate Solutions of the Diffusion Equation 34
3.4.1	Film Model 34
3.4.2	Penetration Model 35
3.4.3	Surface Renewal Model 36

Example 3.1 37
Solution 37

3.5	Physical Interpretation of Some Important Dimensionless Numbers 38
3.5.1	Reynolds Number 38
3.5.2	Prandtl Number and Schmidt Number 39
3.5.3	Nusselt Number 41
3.5.4	Sherwood Number 42
3.5.5	Dimensionless Numbers Commonly Used in Heat and Mass Transfer 44

Example 3.2 44
Solution 44

3.6	Dimensional Analysis 47
3.6.1	Principle of Similitude and Dimensional Homogeneity 47
3.6.2	Finding Dimensionless Numbers and Pi Theorem 48

References 51

4	**Mass Transfer in a Laminar Boundary Layer** 53
4.1	Velocity Boundary Layer 53
4.1.1	Boundary Layer Equation 53
4.1.2	Similarity Transformation 55
4.1.3	Integral Form of the Boundary Layer Equation 57
4.1.4	Friction Factor 58
4.2	Temperature and Concentration Boundary Layers 59
4.2.1	Temperature and Concentration Boundary Layer Equations 59
4.2.2	Integral Form of Thermal and Concentration Boundary Layer Equations 60

Example 4.1 61
Solution 61

4.3	Numerical Solutions of the Boundary Layer Equations	62
4.3.1	Quasi-Linearization Method	62
4.3.2	Correlation of Heat and Mass Transfer Rates	64
	Example 4.2	66
	Solution	66
4.4	Mass and Heat Transfer in Extreme Cases	67
4.4.1	Approximate Solutions for Mass Transfer in the Case of Extremely Large Schmidt Numbers	67
4.4.2	Approximate Solutions for Heat Transfer in the Case of Extremely Small Prandtl Numbers	69
4.5	Effect of an Inactive Entrance Region on Rates of Mass Transfer	70
4.5.1	Polynomial Approximation of Velocity Profiles and Thickness of the Velocity Boundary Layer	70
4.5.2	Polynomial Approximation of Concentration Profiles and Thickness of the Concentration Boundary Layer	71
4.6	Absorption of Gases by a Falling Liquid Film	73
4.6.1	Velocity Distribution in a Falling Thin Liquid Film According to Nusselt	73
4.6.2	Gas Absorption for Short Contact Times	75
4.6.3	Gas Absorption for Long Exposure Times	76
	Example 4.3	77
	Solution	78
4.7	Dissolution of a Solid Wall by a Falling Liquid Film	78
4.8	High Mass Flux Effect in Heat and Mass Transfer in Laminar Boundary Layers	80
4.8.1	High Mass Flux Effect	80
4.8.2	Mickley's Film Model Approach to the High Mass Flux Effect	81
4.8.3	Correlation of High Mass Flux Effect for Heat and Mass Transfer	83
	Example 4.4	86
	Solution	86
	References	87

5	**Heat and Mass Transfer in a Laminar Flow inside a Circular Pipe**	**89**
5.1	Velocity Distribution in a Laminar Flow inside a Circular Pipe	89
5.2	Graetz Numbers for Heat and Mass Transfer	90
5.2.1	Energy Balance over a Small Volume Element of a Pipe	90
5.2.2	Material Balance over a Small Volume Element of a Pipe	92
5.3	Heat and Mass Transfer near the Entrance Region of a Circular Pipe	93
5.3.1	Heat Transfer near the Entrance Region at Constant Wall Temperature	93
5.3.2	Mass Transfer near the Entrance Region at Constant Wall Concentration	94
5.4	Heat and Mass Transfer in a Fully Developed Laminar Flow inside a Circular Pipe	95
5.4.1	Heat Transfer at Constant Wall Temperature	95

5.4.2   Mass Transfer at Constant Wall Concentration   96
5.5   Mass Transfer in Wetted-Wall Columns   97
        Example 5.1   98
        Solution   98
        References   100

6   Motion, Heat and Mass Transfer of Particles   101
6.1   Creeping Flow around a Spherical Particle   101
6.2   Motion of Spherical Particles in a Fluid   104
6.2.1   Numerical Solution of the Drag Coefficients of a Spherical Particle in the Intermediate Reynolds Number Range   104
6.2.2   Correlation of the Drag Coefficients of a Spherical Particle   105
6.2.3   Terminal Velocity of a Particle   106
        Example 6.1   107
        Solution   107
6.3   Heat and Mass Transfer of Spherical Particles in a Stationary Fluid   109
6.4   Heat and Mass Transfer of Spherical Particles in a Flow Field   111
6.4.1   Numerical Approach to Mass Transfer of a Spherical Particle in a Laminar Flow   111
6.4.2   The Ranz–Marshall Correlation and Comparison with Numerical Data   112
        Example 6.2   114
        Solution   114
6.4.3   Liquid-Phase Mass Transfer of a Spherical Particle in Stokes' Flow   115
6.5   Drag Coefficients, Heat and Mass Transfer of a Spheroidal Particle   115
6.6   Heat and Mass Transfer in a Fluidized Bed   117
6.6.1   Void Function   117
6.6.2   Interaction of Two Spherical Particles of the Same Size in a Coaxial Arrangement   117
6.6.3   Simulation of the Void Function   118
        References   120

7   Mass Transfer of Drops and Bubbles   121
7.1   Shapes of Bubbles and Drops   121
7.2   Drag Force of a Bubble or Drop in a Creeping Flow (Hadamard's Flow)   122
7.2.1   Hadamard's Stream Function   122
7.2.2   Drag Coefficients and Terminal Velocities of Small Drops and Bubbles   123
7.2.3   Motion of Small Bubbles in Liquids Containing Traces of Contaminants   126
7.3   Flow around an Evaporating Drop   126
7.3.1   Effect of Mass Injection or Suction on the Flow around a Spherical Particle   126

	7.3.2	Effect of Mass Injection or Suction on Heat and Mass Transfer of a Spherical Particle *128*
		Example 7.1 *129*
		Solution *130*
7.4		Evaporation of Fuel Sprays *131*
7.4.1		Drag Coefficients, Heat and Mass Transfer of an Evaporating Drop *131*
7.4.2		Behavior of an Evaporating Drop Falling Freely in the Gas Phase *132*
		Example 7.2 *134*
		Solution *134*
7.5		Absorption of Gases by Liquid Sprays *136*
		Example 7.3 *137*
		Solution *138*
7.6		Mass Transfer of Small Bubbles or Droplets in Liquids *140*
7.6.1		Continuous-Phase Mass Transfer of Bubbles and Droplets in Hadamard Flow *140*
7.6.2		Dispersed-Phase Mass Transfer of Drops in Hadamard Flow *141*
7.6.3		Mass Transfer of Bubbles or Drops of Intermediate Size in the Liquid Phase *141*
		Example 7.4 *142*
		Solution *142*
		References *143*

8		**Turbulent Transport Phenomena** *145*
8.1		Fundamentals of Turbulent Flow *145*
8.1.1		Turbulent Flow *145*
8.1.2		Reynolds Stress *146*
8.1.3		Eddy Heat Flux and Diffusional Flux *147*
8.1.4		Eddy Transport Properties *148*
8.1.5		Mixing Length Model *149*
8.2		Velocity Distribution in a Turbulent Flow inside a Circular Pipe and Friction Factors *150*
8.2.1		$1/n$-th Power Law *150*
8.2.2		Universal Velocity Distribution Law for Turbulent Flow inside a Circular Pipe *151*
8.2.3		Friction Factors for Turbulent Flow inside a Smooth Circular Pipe *153*
		Example 8.1 *155*
		Solution *155*
8.3		Analogy between Momentum, Heat, and Mass Transfer *156*
8.3.1		Reynolds Analogy *157*
8.3.2		Chilton–Colburn Analogy *158*
		Example 8.2 *160*
		Solution *160*
8.3.3		Von Ka'rman Analogy *161*
8.3.4		Deissler Analogy *162*
		Example 8.3 *164*

Solution   *164*

8.4   Friction Factor, Heat, and Mass Transfer in a Turbulent Boundary Layer   *168*
8.4.1   Velocity Distribution in a Turbulent Boundary Layer   *168*
8.4.2   Friction Factor   *169*
8.4.3   Heat and Mass Transfer in a Turbulent Boundary Layer   *171*
8.5   Turbulent Boundary Layer with Surface Mass Injection or Suction   *172*
Example 8.4   *173*
Solution   *174*
References   *175*

9   Evaporation and Condensation   *177*

9.1   Characteristics of Simultaneous Heat and Mass Transfer   *177*
9.1.1   Mass Transfer with Phase Change   *177*
9.1.2   Surface Temperatures in Simultaneous Heat and Mass Transfer   *178*
9.2   Wet-Bulb Temperatures and Psychrometric Ratios   *179*
Example 9.1   *181*
Solution   *181*
Example 9.2   *182*
Solution   *182*
9.3   Film Condensation of Pure Vapors   *183*
9.3.1   Nusselt's Model for Film Condensation of Pure Vapors   *183*
9.3.2   Effect of Variable Physical Properties   *187*
Example 9.3   *187*
Solution   *188*
9.4   Condensation of Binary Vapor Mixtures   *189*
9.4.1   Total and Partial Condensation   *189*
9.4.2   Characteristics of the Total Condensation of Binary Vapor Mixtures   *190*
9.4.3   Rate of Condensation of Binary Vapors under Total Condensation   *191*
9.5   Condensation of Vapors in the Presence of a Non-Condensable Gas   *192*
9.5.1   Accumulation of a Non-Condensable Gas near the Interface   *192*
9.5.2   Calculation of Heat and Mass Transfer   *193*
9.5.3   Experimental Approach to the Effect of a Non-Condensable Gas   *194*
Example 9.4   *195*
Solution   *196*
9.6   Condensation of Vapors on a Circular Cylinder   *200*
9.6.1   Condensation of Pure Vapors on a Horizontal Cylinder   *200*
9.6.2   Heat and Mass Transfer in the case of a Cylinder with Surface Mass Injection or Suction   *201*
9.6.3   Calculation of the Rates of Condensation of Vapors on a Horizontal Tube in the Presence of a Non-Condensable Gas   *203*
Example 9.5   *204*
Solution   *204*
References   *208*

10 Mass Transfer in Distillation *209*

10.1 Classical Approaches to Distillation and their Paradox *209*
10.1.1 Tray Towers and Packed Columns *209*
10.1.2 Tray Efficiencies in Distillation Columns *210*
10.1.3 HTU as a Measure of Mass Transfer in Packed Distillation Columns *211*
10.1.4 Paradox in Tray Efficiency and HTU *212*
Example 10.1 *214*
Solution *214*
10.2 Characteristics of Heat and Mass Transfer in Distillation *216*
10.2.1 Physical Picture of Heat and Mass Transfer in Distillation *216*
10.2.2 Rate-Controlling Process in Distillation *217*
10.2.3 Effect of Partial Condensation of Vapors on the Rates of Mass Transfer in Binary Distillation *218*
10.2.4 Dissimilarity of Mass Transfer in Gas Absorption and Distillation *221*
Example 10.2 *222*
Solution *222*
Example 10.3 *222*
Solution *222*
10.3 Simultaneous Heat and Mass Transfer Model for Packed Distillation Columns *225*
10.3.1 Wetted Area of Packings *225*
10.3.2 Apparent End Effect *227*
10.3.3 Correlation of the Vapor-Phase Diffusional Fluxes in Binary Distillation *228*
10.3.4 Correlation of Vapor-Phase Diffusional Fluxes in Ternary Distillation *230*
10.3.5 Simulation of Separation Performance in Ternary Distillation on a Packed Column under Total Reflux Conditions *231*
Example 10.4 *233*
Solution *233*
Example 10.5 *239*
Solution *239*
10.4 Calculation of Ternary Distillations on Packed Columns under Finite Reflux Ratio *239*
10.4.1 Material Balance for the Column *239*
10.4.2 Convergence of Terminal Composition *242*
Example 10.6 *244*
Solution *244*
10.5 Cryogenic Distillation of Air on Packed Columns *249*
10.5.1 Air Separation Plant *249*
10.5.2 Mass and Diffusional Fluxes in Cryogenic Distillation *249*
10.5.3 Simulation of Separation Performance of a Pilot-Plant-Scale Air Separation Plant *251*
10.6 Industrial Separation of Oxygen-18 by Super Cryogenic Distillation *252*

10.6.1　Oxygen-18 as Raw Material for PET Diagnostics *252*
10.6.2　A New Process for Direct Separation of Oxygen-18 from Natural Oxygen *253*
10.6.3　Construction and Operation of the Plant *255*
References *257*

Subject Index *271*

Preface

The transfer of materials through interfaces in fluid media is called mass transfer. Mass transfer phenomena are observed throughout Nature and in many fields of industry. Today, fields of application of mass transfer theories have become widespread, from traditional chemical industries to bioscience and environmental industries. The design of new processes, the optimization of existing processes, and solving pollution problems are all heavily dependent on a knowledge of mass transfer.

This book is intended as a textbook on mass transfer for graduate students and for practicing chemical engineers, as well as for academic persons working in the field of mass transfer and related areas. The topics of the book are arranged so as to start from fundamental aspects of the phenomena and then systematically and in a step-by-step way proceed to detailed applications, with due consideration of real separation problems. Important formulae and correlations are clearly described, together with their basic assumptions and the limitations of the theories in practical applications. Comparisons of the theories with numerical solutions or observed data are also provided as far as possible. Each chapter contains some illustrative examples to help readers to understand how to approach actual practical problems.

The book consists of ten chapters. The first three chapters cover the fundamental aspects of mass transfer. The next four chapters deal with laminar mass transfer of various types. The fundamentals of turbulent transport phenomena and mass transfer with phase change are then discussed in two further chapters. The final chapter is a highlight of the book, wherein fundamental principles developed in the previous chapters are applied to real industrial separation processes, and a new model for the design of multi-component distillations on packed columns is proposed, application of which has facilitated the industrial separation of a stable isotope, oxygen-18, by super cryogenic distillation.

Tokyo, July 2006 *Koichi Asano*

Mass Transfer. From Fundamentals to Modern Industrial Applications. Koichi Asano
Copyright © 2006 WILEY-VCH Verlag GmbH & Co. KGaA, Weinheim
ISBN: 3-527-31460-1

1
Introduction

1.1
The Beginnings of Mass Transfer

Separation technology using phase equilibria was perhaps first used by the Greek alchemists of Alexandria [7]. However, modern development of the technology from the viewpoint of rates of mass transfer had to wait until the early 20th century, when W. K. Lewis and W. G. Whitman [4] applied their famous *two-film theory* to gas absorption in 1924. They assumed that there exist two thin fluid films on both sides of an interface, in which the concentration distribution varies sharply and through which transfer of the material takes place by diffusion, and they proposed the important concept of the *mass transfer coefficient* in analogy to the coefficient of heat transfer. Subsequent studies of mass transfer were directed towards experimental approaches to obtaining mass transfer coefficients and delineating empirical correlations thereof. In 1935, R. Higbie [3] applied the transient diffusion model to the absorption of gases by bubbles and proposed a theoretical equation for the prediction of mass transfer coefficients. Although this model represented a milestone in the early days of the studies of mass transfer, its significance was unfortunately not well understood among practical engineers and its application to practical problems was quite limited because it could not deal with mass transfer in flow systems. In 1937, T. K. Sherwood [5] published a well-known textbook on mass transfer, "*Absorption and Extraction*", and demonstrated a systematic approach to the problem.

In 1960, R. B. Bird, W. E. Stewart, and E. N. Lightfoot published a groundbreaking textbook, "*Transport Phenomena*" [1], in which they proposed a new approach to momentum, heat, and mass transfer based on a common understanding that the transport phenomena of these quantities in fluid media are governed by similar fundamental laws and that they should be treated from a common viewpoint in a similar way. In a few decades, this new concept has developed into one of the new fields of engineering science and the title of the book has even become a name of the new engineering science. Nowadays, studies of heat and mass transfer tend to be directed towards a more systematic and theoretical understanding of the phenomena, as opposed to the empirical and case-by-case approach of earlier studies. Many textbooks have since been published in

Mass Transfer. From Fundamentals to Modern Industrial Applications. Koichi Asano
Copyright © 2006 WILEY-VCH Verlag GmbH & Co. KGaA, Weinheim
ISBN: 3-527-31460-1

the field of transport phenomena. Studies of mass transfer are now recognized as a branch of transport phenomena. Although this approach has led to remarkable successes in many fields of practical application, especially in the field of heat transfer, too much emphasis has been placed on systematic interpretation of the phenomena and on the similarity between heat and mass transfer, with consequently too little emphasis on practical applications. As a result, some practically important aspects of mass transfer have inevitably been neglected and because of this comparatively less success has been achieved in this field. The only exceptional case is the textbook, *"Mass Transfer"*, by T. K. Sherwood, R. L. Pigford, and C. R. Wilke, which was published in 1975, but more than 30 years have elapsed since then.

1.2
Characteristics of Mass Transfer

Modern transport phenomena are based on the fundamental assumption that momentum, heat, and mass transfer are similar in nature. However, as far as mass transfer is concerned, there are some specific issues that need to be addressed before any real approach to actual problems can be made. Some of these are summarized in the following.

Phase Equilibria: Figure 1.1 shows the temporal variation in the concentration of a dissolved gas A in a liquid contained in a closed vessel upon contact with the gas at constant pressure and temperature. The rate of increase of the concentration in the liquid is very rapid immediately after exposure to the gas, but it soon becomes gradual and the concentration finally approaches a certain limiting value, which remains constant as long as the pressure and the temperature of the system remain unchanged. This stable state is known as phase equilibrium (saturated solubility of gas A), and the conditions of the phase equilibrium depend solely on the thermodynamic nature of the system. This indicates that the phase equilibrium determines an upper limit for mass transfer, whereas no such limitation exists for heat transfer.

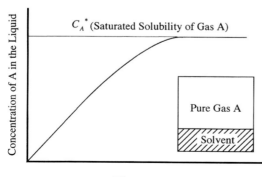

Fig. 1.1 Absorption of a gas by a liquid and solubility of gases in liquids.

Mixture: In momentum and heat transfer, we are mostly concerned with systems consisting of pure fluids, but in mass transfer our main targets are mixed systems of fluids; the simplest case is a binary system, and in most cases we have to deal with ternary or multi-component systems. Consequently, various definitions of concentrations have been used in a case-by-case manner, which can lead to serious confusion in describing rates of mass transfer.

Convective mass flux: Mass transfer can be defined as the transfer of material through an interface between the two phases, whereas diffusion can be defined in terms of the relative motion of molecules from the center of mass of a mixture, moving at the local velocity of the fluid. This means that mass fluxes are not identical to diffusional fluxes at the interface, as is usually assumed in primitive mass transfer models. Rather, mass fluxes are accompanied by convective mass fluxes, as will be discussed in detail in later sections, and can be expressed as the sum of the diffusional fluxes at the interface and the convective mass fluxes. In this respect, mass transfer is completely different from heat transfer, which is not associated with such accompanying fluxes. The existence of convective mass fluxes is a characteristic feature of mass transfer and this needs careful consideration when dealing with mass transfer problems.

High mass flux effect: Convective mass fluxes can also have a significant effect on the velocity and concentration distributions near the interface if the order of magnitude of the flux becomes considerable. This is known as the high mass flux effect.

Effect of latent heat: Mass transfer is a phenomenon involving the transfer of material from one phase to another, and the transfer of material is always accompanied by the energy transfer associated with the phase change, that is, the latent heat. This means that energy transfer is always accompanied by mass transfer, which will affect the interface conditions. In this respect, mass transfer is interrelated with heat transfer through the boundary conditions at the interface. Except for gas absorption in a very low concentration range, the effect is usually quite considerable and we cannot neglect the effect of latent heat.

1.3
Three Fundamental Laws of Transport Phenomena

1.3.1
Newton's Law of Viscosity

Figure 1.2 shows a flow of fluid between two parallel plates, where the upper plate moves at a constant velocity, U [m s^{-1}], and the lower one is at rest. In a steady state, a linear velocity profile is established, as shown in the figure, due to the effect of the viscosity of the fluid. Because of this frictional drag caused by the viscosity of the fluid, a drag force, R_f [N], will act on the surface of the plate. The following empirical equation is known for the fluid friction:

$$\tau = \tau_w = R_f/A \tag{1.1}$$

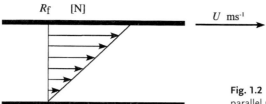

Fig. 1.2 Flow of a viscous fluid between two parallel plates.

$$\tau = -\mu \frac{du}{dy} \tag{1.2}$$

where A is the surface area of the plate [m^2], y is the distance from the wall [m], τ is the shear stress in the fluid [Pa], $\tau_w (\equiv R_f/A)$ is the shear stress at the wall [Pa], and μ is the *viscosity* [Pa s], which is one of the important physical properties of the fluid. Equation (1.2) is usually known as *Newton's law of viscosity*. Fluids are classified into two groups: *Newtonian fluids*, which obey Newton's law of viscosity, and *non-Newtonian fluids*, which do not obey Newton's law. Common fluids such as air, water, and oils generally behave as Newtonian fluids, whereas polymer solutions usually behave as non-Newtonian fluids.

1.3.2
Fourier's Law of Heat Conduction

Figure 1.3 shows the temperature distribution in a solid plate of surface area A [m^2] and thickness δ [m], where the temperature of one surface is kept at T_1 [K] and that of the other at T_2 [K] ($T_1 > T_2$). Heat will be transferred from the hot to the cold surface and this phenomenon is known as the *conduction of heat*. In a steady state, a linear temperature profile is established in the solid and the rate of heat transfer, Q [W], is observed to be proportional to the temperature difference between the two surfaces ($T_1 - T_2$) and the surface area of the plate, A [m^2], and inversely proportional to the thickness of the plate, δ [m]:

$$Q \propto A \frac{(T_1 - T_2)}{\delta} \tag{1.3}$$

The above expression reduces to the following familiar empirical equation as the thickness of the plate approaches an infinitesimally small value:

$$q = \frac{Q}{A} = -\kappa \frac{dT}{dy} \tag{1.4}$$

where $q (\equiv Q/A)$ is the heat flux [W m^{-2}], T is the temperature [K], y [m] is the distance in the direction of heat conduction, and κ is a physical property of the fluid

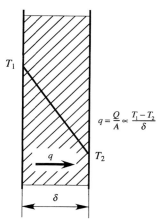

Fig. 1.3 Conduction of heat through a solid wall.

known as the *thermal conductivity* [W m^{-1} K^{-1}]. Equation (1.4) is referred to as *Fourier's law of heat conduction*.

1.3.3
Fick's Law of Diffusion

If we place a small amount of a volatile liquid in the bottom of a test tube and let it be in contact with a dry air stream, as shown in Fig. 1.4, a linear concentration profile is established in the test tube at steady state, and steady evaporation of the liquid will take place. This phenomenon, whereby a transfer of material is caused by a non-uniform distribution of concentration, is called *diffusion*. The following empirical law is known for the rate of diffusion:

$$J_A = -cD \frac{dx_A}{dy} \tag{1.5}$$

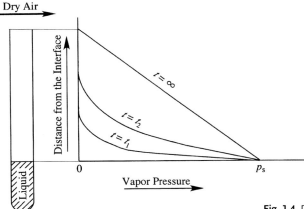

Fig. 1.4 Diffusion of vapor in a gas.

where c is the molar density [kmol m^{-3}], D is the diffusivity [m^{-2} s^{-1}], J_A is the rate of diffusion of component A per unit area of the surface (diffusional flux) [kmol m^{-2} s^{-1}], and y is the distance in the direction of diffusion [m]. Equation (1.5) was first reported by A. Fick [2] in 1855 following observation of the dissolution of salt in water and so is referred to as *Fick's law of diffusion*.

1.4
Summary of Phase Equilibria in Gas-Liquid Systems

The fact that the upper limit of a mass transfer is restricted by the relevant phase equilibrium and that the rate of mass transfer also depends on the phase equilibrium means that we have to be familiar with the phase equilibrium of the system before we can deal with mass transfer problems. Here, we briefly summarize some of the important quantitative relationships of the phase equilibria commonly encountered in gas-liquid systems. More detailed discussions of these topics can be found in standard textbooks on chemical engineering thermodynamics.

Solubility of gases in liquids: The solubility of a gas in a liquid usually increases with increasing pressure and decreases with increasing temperature. For sparingly soluble gases, the well-known Henry's law applies:

$$p_i = H x_i \tag{1.6}$$

where H is the Henry constant [MPa], p_i is the partial pressure of component i [MPa], and x_i is the mole fraction of dissolved gas in the liquid in equilibrium with the gas.

Vapor pressures of pure liquids: The vapor pressure of a pure liquid is a function only of temperature and can be approximated by *Antoine's equation* over a wide range of temperatures:

$$\log p^* = A - \frac{B}{T + C} \tag{1.7}$$

where p^* is the saturated vapor pressure of the liquid [Pa], T [K] is the temperature, and A, B, and C are so-called Antoine's constants.

Vapor pressures of solutions: The vapor pressure of a component i in a solution consisting of members of the same chemical series, such as a mixture of homologous paraffin hydrocarbons, is expressed by the following equation:

$$p_i = p_i^* x_i \tag{1.8}$$

This equation is referred to as *Raoult's law*. The solubility of the vapor of a hydrocarbon in an oil is usually described by *Raoult's law*.

Solutions can be classified into two groups, *ideal solutions*, which obey Raoult's law, and *non-ideal solutions*, which do not obey Raoult's law. Most actual solutions behave as non-ideal solutions. The vapor pressure of a component i in a non-ideal solution can be expressed in a similar way as for an ideal solution through the use of an *activity coefficient*:

$$p_i = p_i^* \gamma_i x_i \tag{1.9}$$

where p_i is the vapor pressure of component i [Pa], p_i^* is the vapor pressure of the pure component i [Pa], x_i is the mole fraction of component i in the liquid [–], and γ_i is the activity coefficient of component i [–].

The estimation of activity coefficients is one of the important subjects of chemical engineering thermodynamics. Readers interested in this subject may refer to the appropriate standard textbooks.

References

1 R. B. Bird, W. E. Stewart, and E. N. Lightfoot, "*Transport Phenomena*", Wiley, (1960).

2 A. Fick, "Ueber Diffusion", *Annalen der Physik und Chemie*, **94**, 59–86 (1855).

3 R. Higbie, "The Rate of Absorption of a Pure Gas into a Still Liquid during Short Periods of Exposure", *Transactions of the American Institute of Chemical Engineers*, **31**, 365–389 (1935).

4 W. K. Lewis and W. G. Whitman, "Principles of Absorption", *Industrial and Engineering Chemistry*, **16**, [12], 1215–1220 (1924).

5 T. K. Sherwood, "*Absorption and Extraction*", McGraw-Hill (1937).

6 T. K. Sherwood, R. L. Pigford, and C. R. Wilke, "*Mass Transfer*", McGraw-Hill (1975).

7 A. J. V. Underwood, "The Historical Development of Distilling Plant", *Transactions of Institutions of Chemical Engineers*, **13**, 34–62 (1935).

2
Diffusion and Mass Transfer

2.1
Motion of Molecules and Diffusion

2.1.1
Diffusion Phenomena

The phenomenon of diffusion is a result of the motion of molecules in a fluid medium. If we observe the motion of molecules in a fluid medium from the viewpoint of the molecular scale, the molecules are seen to be moving randomly in various directions and at various velocities. Here, for the sake of simplicity, we will assume that the motion of molecules is one-dimensional and that the velocities of molecules of the same species i, v_{mi} [m s^{-1}], are the same for all molecules. We further assume that the number density of species i, n_i [molecules m^{-3}], is a function of only the coordinate x.

Figure 2.1 shows a schematic picture of diffusion in a fluid medium on the molecular scale. Let us consider the effect of the motion of molecules of species A in the plane $(x + l)$ and that in the plane $(x - l)$ on the rate of change of the number density of species A in the plane x, where l [m] is the mean free path of species A. The diffusional flux of species A, J_A^* [molecules m^{-2} s^{-1}], may be related to the net number of molecules of species A passing through the plane x per unit area per unit time, that is, the sum of the number of molecules of species A passing through the plane x in the positive direction and the number travelling in the negative direction.

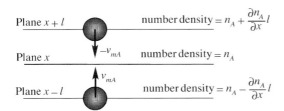

Fig. 2.1 Motion of molecules A and variation of number densities at plane x.

Mass Transfer. From Fundamentals to Modern Industrial Applications. Koichi Asano
Copyright © 2006 WILEY-VCH Verlag GmbH & Co. KGaA, Weinheim
ISBN: 3-527-31460-1

2 Diffusion and Mass Transfer

Therefore, the diffusional flux of species A can be expressed as follows:

$$J_A^* \approx v_{mA} n_A |_{x-l} - v_{mA} n_A |_{x+l}$$

$$= v_{mA}\left(n_A - \frac{\partial n_A}{\partial x}l\right) - v_{mA}\left(n_A + \frac{\partial n_A}{\partial x}l\right) = -(2v_{mA}l)\left(\frac{\partial n_A}{\partial x}\right) \quad (2.1)$$

Equation (2.1), which is obtained from molecular interpretation of the diffusion phenomenon, is mathematically similar to an empirical law of diffusion, namely Fick's law.

2.1.2
Definition of Diffusional Flux and Reference Velocity of Diffusion [1, 2]

If we assume that the motion of the molecules is one-dimensional and that the velocities of molecules of species i, v_i [m s^{-1}], are all the same, we obtain the following equation.

$$v_A = v_B = v_C \ldots = v \quad (2.2)$$

Since the distances between the molecules remain unchanged in this special case, the concentration of each species will remain unchanged. In other words, the diffusion phenomenon is not observed in this case. On the other hand, if the velocities of the molecules of species i, v_i, are not all equal, the distances between the molecules will change with time, as shown in Fig. 2.2. In this case, the concentrations of each component will change with time and the diffusion phenomenon is observed.

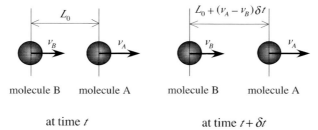

Fig. 2.2 Variation in the distance between two molecules of different velocities with time in a binary system.

If we define the local *molar average velocity* of a mixture of fluids in a small volume element, v^* [m s^{-1}], as:

$$v^* = \sum_{i=1}^{N} x_i \cdot v_i \quad (2.3)$$

and observe the motion of the molecules of species *i* from a coordinate system moving at the same velocity as the molar average velocity, v^*, then the molecules of species *i* are observed to move at a relative velocity of $(v_i - v^*)$. The diffusional flux of species *i* may be related to the relative motion of the molecules of species *i* with respect to the moving coordinate system and can be expressed by the following equation:

$$J_i^* \equiv c_i(v_i - v^*) = cx_i(v_i - v^*) \tag{2.4}$$

Here, c is the molar density of the fluid mixture [kmol m^{-3}], c_i is the partial molar density of component *i* [kmol m^{-3}], and x_i is the mole fraction of component *i*. The diffusional flux defined by Eq. (2.4) is called the *molar diffusional flux*. The velocity v^*, which is the basis for definition of the diffusional flux in Eq. (2.4), is called the *reference velocity of diffusion*. Various definitions of diffusional flux are made possible through different definitions of the reference velocity.

If we sum all of the diffusional fluxes of the system, we obtain the following identical equation:

$$\sum_{i=1}^{N} J_i^* = c \sum_{i=1}^{N} x_i(v_i - v^*) = c \left(\sum_{i=1}^{N} x_i v_i - v^* \sum_{i=1}^{N} x_i \right) = c(v^* - v^*) \equiv 0$$

The above equation indicates an important theorem on diffusional flux in that *the sum of all the diffusional fluxes is always zero*.

$$\sum_{i=1}^{N} J_i^* \equiv 0 \tag{2.5}$$

If we define the *mass average velocity* of the system, or the *baricentric velocity*, as:

$$v = \sum_{i=1}^{N} \omega_i v_i \tag{2.6}$$

we can define another important diffusional flux, namely the *mass diffusional flux* of component *i*, J_i [kg m^{-2} s^{-1}], by using the mass average velocity, v, as the reference velocity.

$$J_i \equiv \rho_i(v_i - v) = \rho \omega_i(v_i - v) \tag{2.7}$$

Here, ρ is the density of the fluid [kg m^{-3}], ρ_i is the partial density of the component *i* [kg m^{-3}], and ω_i is the mass fraction of component *i* [–].

If we sum all the mass diffusional fluxes of Eq. (2.7), we obtain a similar relationship as Eq. (2.5).

$$\sum_{i=1}^{N} J_i \equiv 0 \tag{2.8}$$

The summation theorem for the molar diffusional flux also holds for the mass diffusional flux, J_i.

2.1.3
Binary Diffusion Flux

The detailed and rigorous discussions on binary diffusion in the standard reference [2] indicate that the following equation applies:

$$J_A^* \equiv c x_A (v_A - v^*) = -cD \left(\frac{\partial x_A}{\partial y} \right) \tag{2.9}$$

This implies that the molar diffusional flux defined by Eq. (2.4) reduces to the empirical *Fick's law*. Here, D [m^2 s^{-1}] is the binary diffusion coefficient of component A through component B and y [m] is the distance in the direction of diffusion.

Rearranging Eq. (2.9) by invoking the following relationship for the mole fraction in a binary system:

$$x_A + x_B = 1 \tag{2.10}$$

we obtain the following equation:

$$J_A^* \equiv c x_A (v_A - v^*) = c x_A x_B (v_A - v_B) = c x_B (v^* - v_B) = -J_B^* \tag{2.11}$$

Equation (2.11) indicates that the magnitude of the molar diffusional flux for component A is identical to that for component B, whereas their directions are opposite to one another. Figure 2.3 shows the behavior of each component in binary diffusion.

Rearranging Eq. (2.7) by the use of a similar relationship for the mass fraction:

$$\omega_A + \omega_B = 1 \tag{2.12}$$

we obtain a relationship similar to Eq. (2.11):

$$J_A \equiv \rho \omega_A (v_A - v) = \rho \omega_A \omega_B (v_A - v_B) = \rho \omega_B (v - v_B) = -J_B \tag{2.13}$$

By eliminating $(v_A - v_B)$ from Eqs. (2.11) and (2.13) and subsequently rearranging the resultant equation by the use of the following relationship between the mole fraction and the mass fraction for a binary system:

$$x_A = \frac{\omega_A / M_A}{\omega_A / M_A + \omega_B / M_B} = \frac{(M_B / M_A) \omega_A}{1 + \{(M_B / M_A) - 1\} \omega_A} \tag{2.14}$$

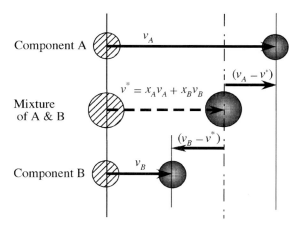

Fig. 2.3 Behavior of diffusion fluxes of each component in binary diffusion.

the following equation is easily obtained:

$$J_A = \frac{\rho \omega_A \omega_B}{c x_A x_B} J_A^* = \frac{M_A M_B}{\overline{M}} \left(-cD \frac{\partial x_A}{\partial y} \right)$$

$$= -cD \left(\frac{M_A M_B}{\overline{M}} \right) \left(\frac{\overline{M}^2}{M_A M_B} \right) \left(\frac{\partial \omega_A}{\partial y} \right) = -\rho D \left(\frac{\partial \omega_A}{\partial y} \right) \quad (2.15)$$

where M_A and M_B are the molecular weights of species A and B [kg kmol^{-1}], respectively, and $\overline{M}(= x_A M_A + x_B M_B)$ is the average molecular weight [kg kmol^{-1}]. Equation (2.15) indicates that the mass diffusional flux, J_A, reduces to the empirical *Fick's law*.

Rearrangement of Eq. (2.15) leads to the following equation:

$$J_A = \frac{1}{1 + x_A (M_A/M_B - 1)} (M_A J_A^*) \quad (2.16)$$

Equation (2.16) indicates that the relationship between the two diffusional fluxes, the molar diffusional flux J_A^* and the mass diffusional flux J_A, is affected not only by the differences in molecular weight but also by concentrations. This fact poses us a fundamental question: "Which diffusional flux, the molar or the mass diffusional flux, is suitable for describing actual mass transfer problems in fluid media?". The conclusion is that either of them will suffice if we are only concerned with mass transfer in a stationary fluid. However, if we are to deal with transport phenomena in a flow system, we have to use the mass diffusional flux since the equation of motion and the energy equation are described in terms of the mass average velocity, v, and not in terms of the molar average velocity, v^*.

This means that if we use molar diffusional flux in a flow system, the difference between the molar and mass average velocities, $(v^* - v)$, will cause a serious error in dealing with the diffusion equation, the details of which will be discussed in Section 3.2.3. For this reason, we will use the mass diffusional flux in the later discussions.

2.2
Diffusion Coefficients

2.2.1
Binary Diffusion Coefficients in the Gas Phase

As stated in the previous section, the diffusional fluxes in a binary system can be described by the following equations:

$$J_A^* = c_i(v_i - v^*) = -cD_{AB}\left(\frac{\partial x_A}{\partial y}\right) \tag{2.9}$$

$$J_A = \rho \omega_i(v_A - v) = -\rho D_{AB}\left(\frac{\partial \omega_A}{\partial y}\right) \tag{2.15}$$

where D_{AB} is the binary diffusion coefficient of component A through component B.

J. O. Hirschfelder et al. [2] derived the following theoretical equation for the binary diffusion coefficients from *the kinetic theory of gases*:

$$D_{AB} = \frac{1.858 \times 10^{-7} T^{3/2} (1/M_A + 1/M_B)^{1/2}}{(P/101325)\sigma_{AB}^2 \Omega(T_D^*)} \quad [\text{m}^2\,\text{s}^{-1}] \tag{2.17}$$

$$\sigma_{AB} = (\sigma_A + \sigma_B)/2$$

$$k/\varepsilon_{AB} = \sqrt{(k/\varepsilon_A)(k/\varepsilon_B)}$$

$$T_D^* = kT/\varepsilon_{AB}$$

Here, M_A and M_B are the molecular weights of components A and B, respectively, P is the pressure [Pa], σ_A and σ_B are the collision diameters of components A and B [Å], ε_A/k [K] and ε_B/k [K] are the characteristic energies of components A and B, respectively, divided by Boltzmann's constant k, and $\Omega(T_D^*)$ [–] is the collision integral, the details of which can be found elsewhere [2, 5].

2.2.2
Multicomponent Diffusion Coefficients in the Gas Phase

According to *the kinetic theory of gases*, the diffusional fluxes in an *N*-component system can be expressed by the following equations:

$$J_1^* = D_{11}\nabla x_1 + D_{12}\nabla x_2 + D_{13}\nabla x_3 + \cdots + D_{1N}\nabla x_N$$

$$J_2^* = D_{21}\nabla x_1 + D_{22}\nabla x_2 + D_{23}\nabla x_3 + \cdots + D_{2N}\nabla x_N$$

$$J_3^* = D_{31}\nabla x_1 + D_{32}\nabla x_2 + D_{33}\nabla x_3 + \cdots + D_{3N}\nabla x_N$$

...

...

$$J_N^* = D_{N1}\nabla x_1 + D_{N2}\nabla x_2 + D_{N3}\nabla x_3 + \cdots + D_{NN}\nabla x_N \tag{2.18}$$

Here, J_i^* is the molar diffusional flux of component *i* [kmol m^{-2} s^{-1}], ∇x_i is the gradient of mole fraction of component *i*, and the coefficients, $D_{11}, D_{12}, \ldots, D_{NN}$, are the multicomponent diffusion coefficients [m^2 s^{-1}]. Although Eq. (2.18) is theoretically derived from the kinetic theory of gases, no reliable method is yet known for estimating the coefficient, D_{ij}, nor is any observed value of D_{ij} known, not even for the simplest multicomponent system, the ternary system. Thus, we cannot use Eq. (2.18) to predict numerical values of diffusional fluxes.

C. R. Wilke [7] proposed the approximate but practically important concept of *effective diffusion coefficient*, which enables us to estimate rates of diffusion in multicomponent systems from the well-established binary diffusion coefficients. The relevant expressions are as follows:

$$J_A^* \equiv cD_{m,A}\nabla x_A \tag{2.19}$$

$$D_{m,A} = \frac{1 - y_A}{\dfrac{y_B}{D_{AB}} + \dfrac{y_C}{D_{AC}} + \dfrac{y_D}{D_{AD}}\cdots} \tag{2.20}$$

where $D_{m,A}$ is the effective diffusion coefficient of component *A* [m^2 s^{-1}] and D_{AB}, D_{AC}, D_{AD}, ... are the binary diffusion coefficients of component *A* through components *B*, *C*, and *D*, respectively [m^2 s^{-1}]. The advantage of this method is that we can easily estimate the diffusional flux of component *i* in a multicomponent system from just the concentration gradient of component *i*, in a similar way as in the case of binary diffusion:

(Diffusion Flux) = (Effective Diffusion Coefficient)(Concentration Gradient)

In Chapter 10, we show how the separation performance of a multicomponent distillation may be predicted by applying the concept of effective diffusion coefficients.

Example 2.1
Calculate the binary diffusion coefficients of water vapor in air at 298.15 K and 1 atm.

Solution
The following parameters are given in standard references:

Air: $M_{Air} = 28.97$, $\sigma_{Air} = 3.62$ Å, $(\varepsilon_{Air}/k) = 97$ K
Water vapor: $M_{water} = 18.02$, $\sigma_{water} = 2.65$ Å, $(\varepsilon_{water}/k) = 356$ K
$\sigma_{AB} = (3.62 + 2.65)/2 = 3.14$
$\varepsilon_{AB}/k = \sqrt{(97)(356)} = 185.8$
$T_D^* = (298.15)/(185.8) = 1.604$

An estimate of the collision integral under these conditions is given by:

$$\Omega(T_D^*) = \frac{1.06036}{(T_D^*)^{0.1561}} + \frac{0.19300}{\exp(0.47635 T_D^*)} + \frac{1.03587}{\exp(1.52996 T_D^*)} + \frac{1.76474}{\exp(3.89411 T_D^*)}$$

$$= 1.167$$

Substitution of the above values into Eq. (2.17) gives:

$$D_{AB} = \frac{1.858 \times 10^{-7} (298.15)^{1.5} \sqrt{1/(28.97) + 1/(18.02)}}{(101325/101325)(3.14)^2 (1.167)} = 2.50 \times 10^{-5} \, m^2 \, s^{-1}$$

The observed value under the same conditions is 2.56×10^{-5} m^2 s^{-1}, which is about 2% larger than the estimated value.

2.3
Rates of Mass Transfer

2.3.1
Definition of Mass Flux

Transfer of a material through a fluid-fluid interface or a fluid-solid interface is called *mass transfer*. If the concentration near the interface is not uniform, mass transfer takes place due to the effect of diffusion. Thus, mass transfer is closely related to diffusion. In this section, we discuss the relationship between mass flux and diffusional flux.

The rate of mass transfer of component *i* is usually expressed as the mass of component *i* passing through unit area of the interface per unit time, which is referred to as the *mass flux* of component *i*, N_i [kg m^{-2} s^{-1}]. Thus,

$$N_i \equiv \rho_s \omega_{is} v_{is} \tag{2.21}$$

2.3 Rates of Mass Transfer

where ρ_s is the density of the fluid at the interface [kg m^{-3}], ω_{is} is the mass fraction of component i at the interface [–], and v_{is} is the velocity of component i at the interface [m s^{-1}].

Rearranging the above equation by the use of Eq. (2.7), that is, the definition of mass diffusional flux.

$$J_i \equiv \rho \omega_i (v_i - v) \qquad (2.7)$$

we obtain the following general equation for the mass flux:

$$N_i = \rho_s \omega_{is} (v_i - v_s) + \rho_s v_s \omega_{is} = J_{is} + \rho_s v_s \omega_{is} \qquad (2.22)$$

The first term on the right-hand side of Eq. (2.22) is the mass diffusional flux at the interface and the second term represents the transfer of component i by the accompanying flow due to diffusional flux at the interface, which is called the *convective mass flux*.

The mass flux can thus be expressed as follows:

(Mass Flux) = (Diffusional Flux) + (Convective Mass Flux)

The fact that the mass flux is always accompanied by convective mass flux is a phenomenon characteristic to mass transfer and has no parallels in momentum or heat transfer. We will discuss the role of convective mass flux below.

From Eq. (2.22) and the summation theorem for mass diffusional flux, Eq. (2.8), the following equation is obtained:

$$\sum_{i=1}^{N} N_i = \sum_{i=1}^{N} (J_{is} + \rho_s v_s \omega_{is}) = \sum_{i=1}^{N} J_{is} + \rho_s v_s \sum_{i=1}^{N} \omega_{is} = \rho_s v_s \qquad (2.23)$$

The above equation is practically important for evaluation of the diffusional flux from the observed mass transfer data for a multicomponent distillation.

The rate of mass transfer in terms of molar units, the *molar flux* of component i, N_i^* [kmol m^{-2} s^{-1}], is defined by the following equation, which can be rearranged into a similar form as Eq. (2.22):

$$N_i^* \equiv N_i/M_i = c_s x_{is} v_{is} = c_s x_{is} (v_{is} - v_s^*) + c_s x_{is} v_s^* = J_{is}^* + c_s x_{is} v_s^* \qquad (2.24)$$

A similar relationship to Eq. (2.23) is obtained for the sum of the molar fluxes:

$$\sum_{i=1}^{N} N_i^* = \sum_{i=1}^{N} J_i^* + c_s v_s^* \sum_{i=1}^{N} x_i = c_s v_s^* \qquad (2.25)$$

Here, c_s [kmol m^{-3}] is the molar density at the interface, x_{is} is the mole fraction of component i at the interface, and v_s^* [m s^{-1}] is the molar average velocity at the interface.

2.3.2
Unidirectional Diffusion in Binary Mass Transfer

A special case of mass transfer, in which component A transfers into a stationary fluid medium of component B, as is the case for the absorption of pure gases by liquids or the evaporation of pure liquids into gases, is called *unidirectional diffusion*.

From the zero mass flux condition for component B, we have the following equation:

$$N_B = J_{Bs} + \rho_s v_s \omega_{Bs} = 0 \tag{2.26}$$

Therefore, the convective velocity at the interface, v_s, is given by:

$$v_s = \frac{-J_B}{\rho_s \omega_{Bs}} = \frac{1}{\rho_s(1-\omega_s)} J_{As} \tag{2.27}$$

Substituting Eq. (2.27) into Eq. (2.22), we obtain the following equation for the mass flux of component A in the case of unidirectional diffusion in a binary system:

$$N_A = J_{As} + \frac{\omega_{As}}{(1-\omega_{As})} J_{As} = \frac{1}{(1-\omega_{As})} J_{As} \tag{2.28}$$

A similar equation is obtained for molar flux for unidirectional diffusion in a binary system:

$$N_A^* = \frac{1}{(1-x_{As})} J_A^* \tag{2.29}$$

For the special case of mass transfer in a multicomponent system in which only component N is stagnant and transfer of the remaining $(N-1)$ components takes place at the interface, the mass flux of component i, N_i [kg m^{-2} s^{-1}], can be expressed by the following equation:

$$N_i = J_{is} + \frac{\omega_{is} \sum\limits_{i \neq N} J_{is}}{1 - \sum\limits_{i \neq N} \omega_{is}} \tag{2.30}$$

2.3.3
Equimolal Counterdiffusion

A special type of binary mass transfer, in which equal numbers of moles of each component are transferred in mutually opposite directions, as is the case with binary distillation, is called *equimolal counterdiffusion*.

From the condition of equimolal counterdiffusion, we have the following equation:

$$\frac{N_A}{M_A} + \frac{N_B}{M_B} = 0 \tag{2.31}$$

Rearranging Eq. (2.31) by the use of Eq. (2.22), we obtain the following equation for the convective flow:

$$v_s = -\frac{(M_B/M_A - 1)}{1 + (M_B/M_A - 1)\omega_{As}}\left(\frac{J_{As}}{\rho_s}\right) \tag{2.32}$$

Substituting Eq. (2.32) into Eq. (2.22), we obtain the following equation:

$$N_A = \frac{J_{As}}{1 + (M_B/M_A - 1)\omega_{As}} \neq J_{As} \tag{2.33}$$

Figure 2.4 shows the relationship between the mass flux and the mass diffusional flux in equimolal counterdiffusion. It can be seen that this relationship is affected not only by the difference in molecular weights but also by the concentrations at the interface.

Although the relationship between the mass flux and the mass diffusional flux in equimolal counterdiffusion is a complicated one, the corresponding relationship for the molar flux is rather simple. We have the following equation:

$$N_A^* \equiv N_A/M_A = J_A^* \; [\text{kmol m}^{-2}\,\text{s}^{-1}] \tag{2.34}$$

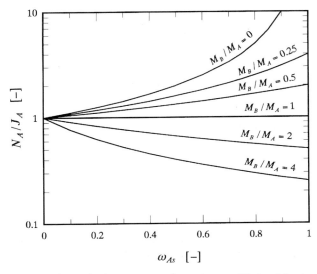

Fig. 2.4 Relationship between mass flux and mass diffusional flux in equimolal counterdiffusion.

since the molar average velocity at the interface, v_s^* [m s^{-1}], is always zero for equimolal counterdiffusion:

$$v_s^* \equiv 0 \tag{2.35}$$

2.3.4
Convective Mass Flux for Mass Transfer in a Mixture of Vapors

In binary and multicomponent distillation, the condensation of mixed vapors and the evaporation of volatile solutions are always accompanied by an interfacial velocity, v_s, caused by condensation of vapors or evaporation of liquids. Although mass transfer in such cases is considerably affected by convective mass flux, this effect has long been neglected by practical engineers. A. Ito and K. Asano [3] presented a theoretical approach to this important problem in binary distillation.

If we consider the energy balance at the interface for a partial condensation of binary vapors, we have the following equation:

$$\lambda_A N_A + \lambda_B N_B + q_G + q_w = 0 \tag{2.36}$$

Rearranging the above equation by the use of Eq. (2.22):

$$N_A = J_{As} + \rho_s v_s \omega_{As}$$
$$N_B = J_{Bs} + \rho_s v_s \omega_{Bs}$$

we obtain the following equation for convective mass flux:

$$\rho_s v_s = \frac{(\lambda_B - \lambda_A) J_{As} - q_G - q_w}{\lambda_A \omega_{As} + \lambda_B \omega_{Bs}} \tag{2.37}$$

Here, λ_A and λ_B are the latent heats of vaporization of components A and B, respectively, at the interface [J kg^{-1}], q_G is the vapor-phase sensible heat flux [W m^{-2}], and q_w is the heat flux from the wall (heat loss) [W m^{-2}].

H. Kosuge and K. Asano [4] also presented a similar approach to convective mass flux in multicomponent distillation:

$$\rho_s v_s = \frac{\sum_{i=1}^{N}(\lambda_N - \lambda_i) J_{is} - q_G - q_w}{\sum_{i=1}^{N} \lambda_i \omega_{is}} \tag{2.38}$$

The significance of Eqs. (2.37) and (2.38), which have permitted a new separation technology for the separation of stable isotopes, will be discussed in Chapter 10.

Example 2.2
Water is placed in a test tube and kept in air at 20 °C and 1 atm. Calculate the rate of evaporation of the water, if the distance between the upper edge of the test tube and the surface of the water is 50 mm.

Solution
We assume that the surface temperature of the water is 20 °C and that the physical properties of the system are as follows:

$D_G = 2.48 \times 10^{-5}$ m^2 s^{-1}, $\rho_G = 1.20$ kg m^{-3}, $\rho_L = 1000$ kg m^{-3}, $\omega_s = 1.43 \times 10^{-3}$ [–].

Since the water is evaporating into air and no transfer of air through the interface takes place, the problem can be regarded as one of unidirectional diffusion. The rate of evaporation of the water can be estimated by applying Eq. (2.28).

$$N_A = \frac{(1.20)(2.48 \times 10^{-5})}{1 - 0.00143} \left(\frac{0.00143}{5 \times 10^{-2}} \right) = 0.852 \times 10^{-6} \text{ kg m}^{-2} \text{ s}^{-1}$$

The rate of decrease of water surface is given by:

$(0.852 \times 10^{-6})(3600)(24)/(1000) = 7.4 \times 10^{-5}$ m day^{-1} = 0.074 mm day^{-1}

2.4 Mass Transfer Coefficients

The discussions in the previous sections have indicated that the rates of mass transfers are closely related to the diffusion at the interface, that is, to the concentration gradients at the interface. In real problems, however, we have no direct means of evaluating concentration gradients at the interface, except in very exceptional cases. In the following, we introduce the important concept of *mass transfer coefficients*, with the aid of which mass transfer rates can be calculated without using the concentration gradients at the interface, as is the case with heat transfer coefficients.

Figure 2.5 shows a schematic representation of the concentration distribution near an interface. Although the variation in the concentration near the interface is very sharp, it becomes more gradual in the region slightly further from the interface and the concentration slowly approaches that in the bulk fluid. Moreover, the concentration at the interface is in phase equilibrium according to the *principle of local equilibrium*. Therefore, the rate of mass transfer can be taken to be proportional to the concentration difference between the interface and the bulk fluid, which is called the *concentration driving force*. If we define the proportionality constant for this case as the *mass transfer coefficient*, the rate of mass transfer can be written by the following equation:

(*Rate of mass transfer*) = (*Mass transfer coefficient*)(*Concentration driving force*)

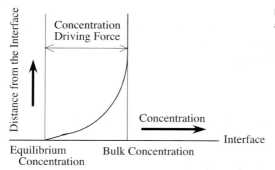

Fig. 2.5 Concentration driving force and mass transfer coefficient.

Although the use of mass transfer coefficients enables us to estimate rates of mass transfer in actual problems, we are faced with another problem. In contrast to heat transfer problems, various definitions of concentration are used in a case-by-case way, which leads to various definitions of mass transfer coefficients. Table 2.1.a shows how the various definitions of concentration in a binary system are interrelated. Table 2.1.b shows the corresponding relationships for multicomponent systems.

Table 2.1.a Relationships between concentrations in a binary system.

	mass fraction ω [−]	mole fraction x [−]	absolute humidity $H\left[\dfrac{\text{kg}-\text{A}}{\text{kg}-\text{B}}\right]$	mole ratio $X\left[\dfrac{\text{mol}-\text{A}}{\text{mol}-\text{B}}\right]$
mass fraction ω [−]	ω	$\dfrac{x}{\dfrac{M_B}{M_A}+\left\{1-\dfrac{M_B}{M_A}\right\}x}$	$\dfrac{H}{1+H}$	$\dfrac{X}{\dfrac{M_B}{M_A}+X}$
mole fraction x [−]	$\dfrac{\omega}{\dfrac{M_A}{M_B}+\left\{1-\dfrac{M_A}{M_B}\right\}\omega}$	x	$\dfrac{H}{\dfrac{M_A}{M_B}+H}$	$\dfrac{X}{1+X}$
absolute humidity $H\left[\dfrac{\text{kg}-\text{A}}{\text{kg}-\text{B}}\right]$	$\dfrac{\omega}{1-\omega}$	$\left(\dfrac{x}{1-x}\right)\left(\dfrac{M_A}{M_B}\right)$	H	$\left(\dfrac{M_A}{M_B}\right)X$
mole ratio $X\left[\dfrac{\text{mol}-\text{A}}{\text{mol}-\text{B}}\right]$	$\left(\dfrac{M_B}{M_A}\right)\left(\dfrac{\omega}{1-\omega}\right)$	$\left(\dfrac{x}{1-x}\right)$	$\left(\dfrac{M_B}{M_A}\right)H$	X

2.4 Mass Transfer Coefficients

Table 2.1.b Relationships between various definitions of concentration in a multicomponent system.

	mass fraction ω_A [–]	mole fraction x_A [–]	partial density ρ_A [kg m^{-3}]	molar density c_A [kmol m^{-3}]	partial pressure p_A [kPa]
mass fraction ω_A [–]	ω	$\dfrac{x_A M_A}{\sum_i x_i M_i}$	$\dfrac{\rho_A}{\sum_i \rho_i}$	$\dfrac{c_A M_A}{\sum_i c_i M_i}$	$\dfrac{p_A M_A}{\sum_i p_i M_i}$
mole fraction x_A [–]	$\dfrac{(\omega_A/M_A)}{\sum_i (\omega_i/M_i)}$	x_A	$\dfrac{\rho_A/M_A}{\sum_i (\rho_i/M_i)}$	$\dfrac{c_A}{\sum_i c_i}$	$\dfrac{p_A}{\sum_i p_i}$
partial density ρ_A [kg m^{-3}]	$\rho \omega_A$	$\dfrac{\rho x_A M_A}{\sum_i x_i M_i}$	ρ_A	$c_A M_A$	$\dfrac{M_A p_A}{RT}$
molar density c_A [kmol m^{-3}]	$\dfrac{\rho \omega_A}{M_A}$	$c x_A$	$\dfrac{\rho_A}{M_A}$	c_A	$\dfrac{p_A}{RT}$
partial pressure p_A [kPA]	$\dfrac{(\omega_A/M_A)P}{\sum_i (\omega_i/M_i)}$	$P x_A$	$\dfrac{RT \rho_A}{M_A}$	$c_A RT$	p_A

Mixture: $\sum_i x_i = 1$, $\sum_i \omega_i = 1$, $\rho = \sum_i \rho_i$, $c = \sum_i c_i$, $P = \sum_i p_i$

$$M = \sum_i M_i x_i = \left(\sum_i \frac{\omega_i}{M_i} \right)^{-1}$$

Table 2.2 summarizes various definitions of mass transfer coefficients.

Table 2.2 Mass transfer coefficients.

Mass transfer coefficient	Unit	Definition	Driving force	Phase
k_y	[kmol m^{-2} s^{-1}]	$N_A^* = k_y (y_s - y_\infty)$	Δy	
k_G	[kmol m^{-2} s^{-1} kPa^{-1}]	$N_A^* = k_G (p_s - p_\infty)$	Δp	
k_Y	[kmol m^{-2} s^{-1}]	$N_A^* = k_Y (Y_s - Y_\infty)$	ΔY	Gas phase
k	[m s^{-1}]	$N_A = \rho_G k (\omega_{Gs} - \omega_{G\infty})$	$\Delta \omega_G$	
k_H	[kg m^{-2} s^{-1}]	$N_A = k_H (H_s - H_\infty)$	ΔH	
k_L	[m s^{-1}]	$N_A^* = k_L (c_s - c_\infty)$	Δc	
k_x	[kmol m^{-2} s^{-1}]	$N_A^* = k_x (x_s - x_\infty)$	Δx	Liquid phase
k_X	[kmol m^{-2} s^{-1}]	$N_A^* = k_X (X_s - X_\infty)$	ΔX	
k	[m s^{-1}]	$N_A = \rho_L k (\omega_{Ls} - \omega_{L\infty})$	$\Delta \omega_L$	

c = molar density [mol m^{-3}], H = absolute humidity [–], M_A = molecular weight [kg kmol^{-1}], N_A = mass flux [kg m^{-2} s^{-1}], $N_A^* = N_A/M_A$ = molar flux [kmol m^{-2} s^{-1}], p = partial pressure [kPa], x, y = mole fraction [–], $X = x/(1-x)$ [–], $Y = y/(1-y)$ [–], ω = mass fraction [–].

2 Diffusion and Mass Transfer

Example 2.3
Show the mutual relationship between the mass transfer coefficients k, k_H, k_c, k_y, and k_G.

Solution
From the definitions of mass transfer coefficients shown in Tab. 2.2, we have the following equations:

$$N_A = \rho k(\omega_s - \omega_\infty) = k_H(H_s - H_\infty) \tag{A}$$

$$N_A/M_A = N_A^* = k_c(c_s - c_\infty) = k_y(y_s - y_\infty) = k_G(p_s - p_\infty) \tag{B}$$

From Tab. 2.1, we also have:

$$H = \frac{\omega}{1-\omega}, \quad c = \frac{p}{RT}, \quad y = \frac{p}{P} = \frac{(M_B/M_A)\omega_A}{1 + \{M_B/M_A - 1\}\omega_A} \tag{C}$$

Substituting these equations into Eq. (A) or (B), we obtain the following equations:

$$k = \frac{k_H}{\rho(1-\omega_s)(1-\omega_\infty)} \tag{D}$$

$$k_y = k_G P = k_c(P/RT)$$
$$= \left(\frac{\rho k}{M_A}\right)\left\{1 + \left(\frac{M_B}{M_A} - 1\right)\omega_s\right\}\left\{1 + \left(\frac{M_B}{M_A} - 1\right)\omega_\infty\right\}\left(\frac{M_A}{M_B}\right)$$
$$\approx ck\left\{1 + \left(\frac{M_B}{M_A}\right)\omega_s\right\} \tag{E}$$

2.5
Overall Mass Transfer Coefficients

In industrial separation processes, there are usually concentration distributions on both sides of the interface, except in very rare cases such as the absorption of pure gases or the evaporation of pure liquids. Figure 2.6 shows a schematic representation of the concentration distribution near the interface in such a situation. Since we have no direct means of evaluating the concentration at the interface, we cannot calculate the rate of mass transfer by direct use of individual mass transfer coefficients. If, however, we define *overall mass transfer coefficients* in analogy to overall heat transfer coefficients, we can easily calculate rates of mass transfer as if the mass transfer resistance were only on one side of the interface.

Taking into account the concentration distribution shown in Fig. 2.6, we have the following relationships:

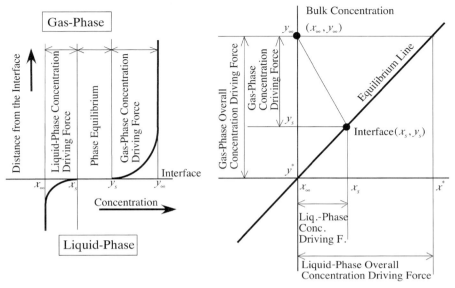

Fig. 2.6 Concentration profiles near the gas–liquid interface and overall mass transfer coefficient.

$$N_A^* = k_y(y_\infty - y_s) = k_x(x_s - x_\infty) = K_y(y_\infty - y^*) = K_x(x^* - x_\infty) \tag{2.39}$$

Here, k_x and k_y are the liquid- and gas-phase mass transfer coefficients, K_x and K_y are the liquid- and gas-phase overall mass transfer coefficients [kmol m^{-2} s^{-1}], N_A^* is the molar flux [kmol m^{-2} s^{-1}], x_s is the equilibrium mole fraction of the liquid at the interface, x_∞ is the mole fraction of the bulk liquid, x^* is the mole fraction of the liquid in equilibrium with the bulk gas, y_s is the equilibrium mole fraction of the gas at the interface, y_∞ is the mole fraction of the bulk gas, and y^* is the mole fraction of the gas in equilibrium with the bulk liquid.

If we assume a linear equilibrium relationship between the gas and the liquid:

$$y^* = mx + b \tag{2.40}$$

we can obtain the following relationship from Fig. 2.6:

$$(y_\infty - y^*) = (y_\infty - y_s) + m(x_s - x_\infty)$$

Substituting the above equation into Eq. (2.39) and rearranging the resultant equation, we will obtain the following equation for the overall mass transfer coefficients:

$$\frac{1}{K_y} = \frac{m}{K_x} = \frac{1}{k_y} + \frac{m}{k_x} \tag{2.41}$$

The significance of Eq. (2.41) is that we can estimate rates of mass transfer without evaluating the surface concentration if we have data for the individual mass transfer coefficients, k_x and k_y.

Although the above discussion is based on a mole fraction concentration driving force, similar results are obtained if we consider mass fraction driving force or mole ratio driving force.

References

1 R. B. Bird, "Advances in Chemical Engineering Vol. 1", p. 156–239, Academic Press (1956).
2 J. O. Hirschfelder, C. F. Curtis, and R. B. Bird, "Molecular Theory of Gases and Liquids", p. 441–610, John Wiley and Sons (1952).
3 A. Ito and K. Asano, "Thermal Effects in Non-Adiabatic Binary Distillation; Effects of Partial Condensation of Mixed Vapors on the Rates of Heat and Mass Transfer and Prediction of H. T. U.", Chemical Engineering Science, 37, [1], 1007–1014 (1983).
4 H. Kosuge and K. Asano, "Mass and Heat Transfer in Ternary Distillation of Methanol-Ethanol-Water Systems by a Wetted-Wall Column", Journal of Chemical Engineering of Japan, 15, [4], 268–273 (1982).
5 R. C. Reid, J. M. Prausnitz, and T. K. Sherwood, "The Properties of Gases and Liquids", 3rd Edition, p. 544–549, McGraw-Hill (1977).
6 T. K. Sherwood, R. L. Pigford, and C. R. Wilke, "Mass Transfer", p. 8–51, 179–180, McGraw-Hill (1975).
7 C. R. Wilke, "Diffusional Properties of Multicomponent Gases", Chemical Engineering Progress, 46, [2], 95–104 (1950).

3
Governing Equations of Mass Transfer

3.1
Laminar and Turbulent Flow

Mass transfer takes place at the interface between two mutually insoluble fluids or at the interface between a fluid and a solid. Since the phenomenon is closely related to fluid flow, we have to understand the fundamental nature of flow before going into details of the phenomenon.

O. Reynolds [8] was the first to address the fundamental aspects of the mechanism of flow. In 1883, he studied the flow of water in a horizontal glass tube by injecting tracer liquid colored with a dye from a small nozzle placed along the center-line of the tube. According to his observations, there are two types of flow, as shown in Figs. 3.1.a and b. The tracer is observed in a form like a single string if the flow rate of the water is relatively low (Fig. 3.1.a), and similar results are observed for tracer injected from different radial positions. This type of flow, in which a fluid behaves as though it were composed of parallel layers, is called *laminar flow*. On the other hand, if the flow rate of the water exceeds a certain critical value, the flow suddenly changes to a completely different type and many eddies are observed, as shown in Fig. 3.1.b. Under these conditions, the tracer will spread in the downstream region of the tube and finally the whole tube section is filled with the dispersed tracer. This type of flow, in which irregular motion of the fluid due to eddies is observed, is called *turbulent flow*. These two types of flow, laminar and turbulent flow, are commonly observed not only for water but also for almost every type of fluid, and the transition from laminar to turbulent flow is always seen if the flow rate exceeds a certain critical value.

Transport phenomena in laminar and turbulent flows are completely different in nature; hence, we cannot apply the same governing equations to describe transport phenomena in laminar and turbulent flows. In this chapter, we describe the governing equations for transport phenomena in laminar flows. Some approximate solutions of the diffusion equation and physical interpretations of some important dimensionless numbers in heat and mass transfer are also discussed. The fundamental nature of turbulent transport phenomena will be discussed in Chapter 8.

Mass Transfer. From Fundamentals to Modern Industrial Applications. Koichi Asano
Copyright © 2006 WILEY-VCH Verlag GmbH & Co. KGaA, Weinheim
ISBN: 3-527-31460-1

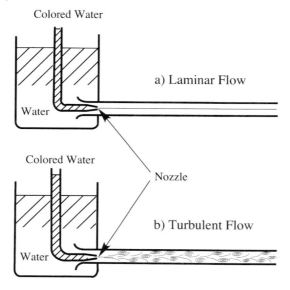

Fig. 3.1 Laminar and turbulent flow; experiment by O. Reynolds.

3.2
Continuity Equation and Diffusion Equation

3.2.1
Continuity Equation

We will assume *the law of conservation of mass* in fluid media. Figure 3.2 shows the conservation of mass for a small volume element $dV(dx, dy, dz)$ at position $P(x, y, z)$ in Cartesian coordinates. The mass balance for the volume element dV can be expressed by the following equation:

(Rate of accumulation of mass in dV)
= (Rate of mass flowing into dV) − (Rate of mass flowing out from dV) (3.1)

The rate of accumulation of fluid mass in dV during the time interval dt is equal to the net increase of fluid mass flowing into dV through three pairs of parallel surface elements perpendicular to the x-, y-, and z-coordinates. The increase of fluid mass flowing into dV through a pair of parallel surface elements perpendicular to the x-coordinate, $dy\,dz$, can be expressed by the following equation:

$$\{\rho u|_x - \rho u|_{x+dx}\}dy dz = \left\{\rho u - \left(\rho u + \frac{\partial \rho u}{\partial x}\right)dx\right\}dy dz = -\left(\frac{\partial \rho u}{\partial x}\right)dV \quad (3.2)$$

Similar relationships are obtained for fluid mass flowing into dV through two pairs of parallel surface elements perpendicular to the y- and z-coordinates. There-

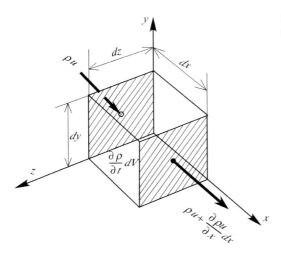

Fig. 3.2 Conservation of mass in a small volume element dxdydz.

fore, the law of conservation of mass for a small volume element $dV (= dx\, dy\, dz)$ can be expressed by the following equation:

$$\frac{\partial \rho}{\partial t} + \frac{\partial \rho u}{\partial x} + \frac{\partial \rho v}{\partial y} + \frac{\partial \rho w}{\partial z} = 0 \tag{3.3}$$

where ρ is the density of the fluid [kg m^{-3}], and u, v, w are components of the velocity (baricentric velocity) of the fluid in the x-, y-, and z-coordinates [m s^{-1}], respectively. Equation (3.3) is referred to as the *continuity equation*.

In the case of *incompressible fluids*, the density of which is constant, as for common liquids, the continuity equation, Eq. (3.3), can be simplified to the following equation:

$$\frac{\partial u}{\partial x} + \frac{\partial v}{\partial y} + \frac{\partial w}{\partial z} = 0 \tag{3.4}$$

3.2.2
Diffusion Equation in Terms of Mass Fraction

In contrast to the equation of motion and the energy equation, which are rigorously derived by application of the conservation laws, the derivation of the diffusion equation is insufficiently described in standard textbooks. Here, we present a rigorous derivation of the diffusion equation by applying *the law of conservation of mass* in fluid media.

The conservation of mass for component A in a small volume element of a fluid mixture gives the following equation, *the continuity equation for component A*:

$$\frac{\partial \rho_A}{\partial t} + \frac{\partial \rho_A u_A}{\partial x} + \frac{\partial \rho_A v_A}{\partial y} + \frac{\partial \rho_A w_A}{\partial z} - r_A M_A = 0 \tag{3.5}$$

Here, ρ_A is the partial density of A [kg m^{-3}], M_A is the molecular weight of A [kg kmol^{-1}], r_A is the rate of production of A by chemical reaction per unit volume of fluid, and u_A, v_A, w_A are components of the velocity of A [m s^{-1}] in the x-, y-, and z-coordinates, respectively. The components of the velocity of A are related to the components of the local mass average velocity through the definition of mass diffusional flux:

$$J_{Ax} \equiv \rho_A (u_A - u) \tag{2.7a}$$

$$J_{Ay} \equiv \rho_A (v_A - v) \tag{2.7b}$$

$$J_{Az} \equiv \rho_A (w_A - w) \tag{2.7c}$$

Rearranging each term of Eq. (3.5) using the appropriate part of Eq. (2.7) and the following equation for the mass fraction:

$$\rho_A = \rho \omega_A \tag{3.6}$$

we have the following equations:

$$\frac{\partial \rho_A}{\partial t} = \frac{\partial \rho \omega_A}{\partial t} = \rho \frac{\partial \omega_A}{\partial t} + \omega_A \frac{\partial \rho}{\partial t}$$

$$\frac{\partial \rho_A u_A}{\partial x} = \frac{\partial \rho u \omega_A}{\partial x} + \frac{\partial}{\partial x} \rho \omega_A (u_A - u) = \rho u \frac{\partial \omega_A}{\partial x} + \omega_A \frac{\partial \rho u}{\partial x} + \frac{\partial J_{Ax}}{\partial x}$$

$$\frac{\partial \rho_A v_A}{\partial y} = \rho v \frac{\partial \omega_A}{\partial y} + \omega_A \frac{\partial \rho v}{\partial y} + \frac{\partial J_{Ay}}{\partial y}$$

$$\frac{\partial \rho_A w_A}{\partial z} = \rho w \frac{\partial \omega_A}{\partial z} + \omega_A \frac{\partial \rho w}{\partial z} + \frac{\partial J_{Aw}}{\partial z}$$

Substituting these equations into Eq. (3.5) and rearranging the resultant equation by use of the continuity equation, Eq. (3.3), we have the following equation:

$$\rho \frac{\partial \omega_A}{\partial t} + \rho u \frac{\partial \omega_A}{\partial x} + \rho v \frac{\partial \omega_A}{\partial y} + \rho w \frac{\partial \omega_A}{\partial z} + \left(\frac{\partial J_{Ax}}{\partial x} + \frac{\partial J_{Ay}}{\partial y} + \frac{\partial J_{Az}}{\partial z} \right) - r_A M_A = 0 \tag{3.7}$$

Equation (3.7) is referred to as the *diffusion equation* in terms of mass fraction.

In the following, we consider the diffusion equation for a binary system of constant physical properties without chemical reaction. In this special case, we can apply *Fick's law* of diffusion:

$$J_{Ax} = -\rho D \frac{\partial \omega_A}{\partial x} \tag{2.15a}$$

$$J_{Ay} = -\rho D \frac{\partial \omega_A}{\partial y} \qquad (2.15b)$$

$$J_{Az} = -\rho D \frac{\partial \omega_A}{\partial z} \qquad (2.15c)$$

Rearranging Eq. (3.7), we obtain the following equation:

$$\frac{\partial \omega_A}{\partial t} + u\frac{\partial \omega_A}{\partial x} + v\frac{\partial \omega_A}{\partial y} + w\frac{\partial \omega_A}{\partial z} = D\left(\frac{\partial^2 \omega_A}{\partial x^2} + \frac{\partial^2 \omega_A}{\partial y^2} + \frac{\partial^2 \omega_A}{\partial z^2}\right) \qquad (3.8)$$

3.2.3
Diffusion Equation in Terms of Mole Fraction

The diffusion equation in terms of mole fraction can also be obtained from *the law of conservation of mass*. Rearranging Eq. (3.7) for the case without chemical reaction ($r_A = 0$) by use of the following equation for the partial density:

$$\rho_A = c_A M_A = c x_A M_A \qquad (3.9)$$

we have the following equation:

$$\left(\frac{\partial x_A}{\partial t} + u_A^*\frac{\partial x_A}{\partial x} + v_A^*\frac{\partial x_A}{\partial y} + w_A^*\frac{\partial x_A}{\partial z}\right) + \frac{1}{c}\left(\frac{\partial J_{Ax}^*}{\partial x} + \frac{\partial J_{Ay}^*}{\partial y} + \frac{\partial J_{Az}^*}{\partial z}\right) = 0 \qquad (3.10)$$

where c is the molar density [kmol m^{-3}], c_A is the molar density of A [kmol m^{-3}], x_A is the mole fraction of A, u^*, v^*, w^* are components of the local molar average velocity in the x-, y-, and z-coordinates, and J_{Ax}^*, J_{Ay}^*, J_{Az}^* are components of the molar diffusional flux [kmol m^{-2} s^{-1}] in the x-, y-, and z-coordinates, respectively.

$$J_{Ax}^* = c x_A (u_A - u^*) \qquad (2.4a)$$

$$J_{Ay}^* = c x_A (v_A - v^*) \qquad (2.4b)$$

$$J_{Az}^* = c x_A (w_A - w^*) \qquad (2.4c)$$

If we further assume a binary system of constant physical properties, then the well-known *Fick's law of diffusion* also applies:

$$J_{Ax}^* = -cD\frac{\partial x_A}{\partial x} \qquad (2.9a)$$

$$J_{Ay}^* = -cD\frac{\partial x_A}{\partial y} \qquad (2.9b)$$

$$J_{Az}^* = -cD\frac{\partial x_A}{\partial z} \tag{2.9c}$$

Substituting the above equations into Eq. (3.10), we obtain the following equation:

$$\left\{\frac{\partial x_A}{\partial t} + u\frac{\partial x_A}{\partial x} + v\frac{\partial x_A}{\partial y} + w\frac{\partial x_A}{\partial z} - D\left(\frac{\partial^2 x_A}{\partial x^2} + \frac{\partial^2 x_A}{\partial y^2} + \frac{\partial^2 x_A}{\partial z^2}\right)\right\}$$
$$+ \left\{(u^* - u)\frac{\partial x_A}{\partial x} + (v^* - v)\frac{\partial x_A}{\partial y} + (w^* - w)\frac{\partial x_A}{\partial z}\right\} = 0 \tag{3.11}$$

Let us consider the order of magnitude of the second term on the left-hand side of Eq. (3.11), the effect of the difference between the two velocities, the molar and the mass average velocity. The component of the velocity difference in the x-coordinate can be expressed by:

$$(u^* - u) = (x_A u_A + x_B u_B) - (\omega_A u_A + \omega_B u_B) = \frac{x_A x_B (M_B - M_A)}{\overline{M}}(u_A - u_B)$$

Similar relationships are obtained for the components in the y- and z-coordinates. These equations indicate that we can neglect the second term on the left-hand side of Eq. (3.11), if either of the following conditions is satisfied:

Case 1) Equal molecular weights of the two components, $M_A = M_B$

Case 2) The concentration of one of the two components is very low,

$$x_A \approx 0 \quad \text{or} \quad x_B = 1 - x_A \approx 0$$

Under these special conditions, Eq. (3.11) can be approximated by the following equation, and a familiar diffusion equation in terms of mole fractions is obtained.

$$\frac{\partial x_A}{\partial t} + u\frac{\partial x_A}{\partial x} + v\frac{\partial x_A}{\partial y} + w\frac{\partial x_A}{\partial z} = D\left(\frac{\partial^2 x_A}{\partial x^2} + \frac{\partial^2 x_A}{\partial y^2} + \frac{\partial^2 x_A}{\partial z^2}\right) \tag{3.12}$$

An important conclusion is that Eq. (3.12) is an approximate one and is only valid in the dilute concentration range. That is to say, we can apply Eq. (3.12) to ordinary gas absorption, where the concentrations are usually very dilute, but we cannot apply the equation to distillation, where the variation of concentration within the apparatus is significant.

3.3
Equation of Motion and Energy Equation

3.3.1
The Equation of Motion (Navier–Stokes Equation)

The equation of motion in a Newtonian fluid, which is called the *Navier–Stokes equation*, can be obtained from the momentum balance in a small volume element after rigorous but tedious calculations.

The equation of motion for a fluid of constant physical properties in Cartesian coordinates can be written as follows:

$$\frac{\partial u}{\partial t} + u\frac{\partial u}{\partial x} + v\frac{\partial u}{\partial y} + w\frac{\partial u}{\partial z} = -\frac{1}{\rho}\frac{\partial P}{\partial x} + \frac{\mu}{\rho}\left(\frac{\partial^2 u}{\partial x^2} + \frac{\partial^2 u}{\partial y^2} + \frac{\partial^2 u}{\partial z^2}\right) + g_x \quad (3.13\text{a})$$

$$\frac{\partial v}{\partial t} + u\frac{\partial v}{\partial x} + v\frac{\partial v}{\partial y} + w\frac{\partial v}{\partial z} = -\frac{1}{\rho}\frac{\partial P}{\partial y} + \frac{\mu}{\rho}\left(\frac{\partial^2 v}{\partial x^2} + \frac{\partial^2 v}{\partial y^2} + \frac{\partial^2 v}{\partial z^2}\right) + g_y \quad (3.13\text{b})$$

$$\frac{\partial w}{\partial t} + u\frac{\partial w}{\partial x} + v\frac{\partial w}{\partial y} + w\frac{\partial w}{\partial z} = -\frac{1}{\rho}\frac{\partial P}{\partial z} + \frac{\mu}{\rho}\left(\frac{\partial^2 w}{\partial x^2} + \frac{\partial^2 w}{\partial y^2} + \frac{\partial^2 w}{\partial z^2}\right) + g_z \quad (3.13\text{c})$$

where g_x, g_y, g_z are the components of gravitational acceleration $\boldsymbol{g}$ [m s^{-2}], P is the pressure [Pa], μ is the viscosity [Pa s], and ρ is the density [kg m^{-3}]. Readers interested in the detailed derivation of the equation of motion may refer to a standard textbook [9].

3.3.2
The Energy Equation

The energy equation is also derived by application of *the law of conservation of energy*. Here, we present the energy equation for a fluid of constant physical properties under moderate flow conditions, where the effect of viscous dissipation is negligibly small.

$$\frac{\partial T}{\partial t} + u\frac{\partial T}{\partial x} + v\frac{\partial T}{\partial y} + w\frac{\partial T}{\partial z} = \frac{\kappa}{\rho c_p}\left(\frac{\partial^2 T}{\partial x^2} + \frac{\partial^2 T}{\partial y^2} + \frac{\partial^2 T}{\partial z^2}\right) \quad (3.14)$$

Readers interested in the exact derivation and a more general form of the energy equation may refer to a standard textbook [9].

3.3.3
Governing Equations in Cylindrical and Spherical Coordinates

We have discussed the nature of the governing equations for transport phenomena in a laminar flow using Cartesian coordinates. The corresponding governing equations in terms of cylindrical and spherical coordinates are obtained through transformation of the coordinates after tedious calculations. The governing equations for these coordinate systems are summarized in Appendix A.

3.4
Some Approximate Solutions of the Diffusion Equation

3.4.1
Film Model [6]

In a special case of mass transfer, as in the evaporation of a pure liquid in a diffusion cell or the absorption of a pure gas in an agitated vessel, transfer of material is one-dimensional and we can neglect the convective terms in the diffusion equation.

$$D\frac{d^2\omega}{dy^2} = 0 \tag{3.15}$$

Integration of Eq. (3.15) gives:

$$D\frac{d\omega}{dy} = \text{constant} \tag{3.16}$$

Equation (3.16) indicates that the concentration distribution is linear, as shown in Fig. 3.3. If we assume *unidirectional diffusion* and that the surface concentration is very low ($\omega_s \approx 0$), the mass flux of component A, N_A [kg m^{-2} s^{-1}], can be expressed by the following equation:

$$N_A = \frac{\rho D}{1-\omega_s}\frac{(\omega_s - \omega_\infty)}{\delta_c} \approx \rho D \frac{(\omega_s - \omega_\infty)}{\delta_c} \tag{3.17}$$

Here, D is the diffusivity [m^2 s^{-1}], δ_c is the thickness of the concentration film in which the concentration profile exists [m], ρ is the density of the fluid [kg m^{-3}], and ω_s and ω_∞ are the mass fractions at the interface and in the bulk fluid, respectively. Equation (3.17) indicates that the rate of mass transfer in this special case is proportional to the diffusion coefficient and inversely proportional to the

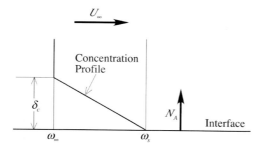

Fig. 3.3 Film model.

thickness of the film, δ_c. This model is commonly known as the *film model*. Although the *film model* offers some explanation of the mechanism of mass transfer in fluid media, it does not provide us with any means of estimating the thickness of the concentration film. Due to this disadvantage, application of the model is restricted to mass transfer in a diffusion cell, in which case the diffusion coefficients can be determined by the use of Eq. (3.17).

3.4.2
Penetration Model

In the absorption of gases from bubbles or absorption by wetted-wall columns, the mass transfer surface is formed instantaneously and transient diffusion of the material takes place. Under these circumstances, we can neglect the convective terms in the diffusion equation and so it reduces to a similar equation as applies in the case of transient heat conduction:

$$\frac{\partial \omega}{\partial t} = D \frac{\partial^2 \omega}{\partial x^2} \tag{3.18}$$

The boundary conditions are:

$$t = 0, \quad x > 0: \quad \omega = \omega_\infty$$
$$t > 0, \quad x = 0: \quad \omega = \omega_s$$

The solutions of the above partial differential equation of heat conduction type, Eq. (3.18), are well described in a standard textbook [3], and the solution for the above boundary conditions is given by the following equation:

$$\frac{\omega_s - \omega}{\omega_s - \omega_\infty} = \mathrm{erf}\left(\frac{x}{2\sqrt{Dt}}\right) \tag{3.19}$$

where erf(x) is the *error function* defined by Eq. (3.20):

$$\mathrm{erf}(x) \equiv \frac{2}{\sqrt{\pi}} \int_0^x \exp(-z^2)\, dz \tag{3.20}$$

Figure 3.4 shows the concentration profile obtained by this model, from which it can be seen that this profile varies with time. Since the process of mass transfer in gas absorption is a unidirectional diffusion and the surface concentration is usually very low ($\omega_s \approx 0$), the mass flux of component A, N_A [kg m^{-2} s^{-1}], can be estimated by means of the following equation:

$$N_A = \frac{-\rho D}{1 - \omega_s}\left(\frac{\partial \omega}{\partial x}\right)_{x=0} \approx -\rho D \left(\frac{\partial \omega}{\partial x}\right)_{x=0} \tag{3.21}$$

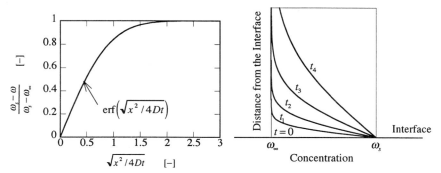

Fig. 3.4 Higbie's penetration model.

Substituting Eq. (3.19) into Eq. (3.21), the rate of mass transfer at time t is given by the following equation:

$$N_A(t) = \rho \sqrt{\frac{D}{\pi t}} (\omega_s - \omega_\infty) \tag{3.22}$$

The mass transfer coefficient is given by:

$$k(t) = \sqrt{\frac{D}{\pi t}} \tag{3.23}$$

The average mass transfer coefficient during a time interval t_c [s] is obtained by integrating Eq. (3.23) as:

$$\bar{k} = \frac{1}{t_c} \int_0^{t_c} k(t)\, dt = 2\sqrt{\frac{D}{\pi t_c}} \tag{3.24}$$

Equations (3.23) and (3.24) were first proposed by R. Higbie [5] in 1935, and the model is called *Higbie's penetration model*.

3.4.3
Surface Renewal Model

P. V. Danckwerts [4] modified *Higbie's penetration model* for mass transfer in the liquid phase, on the assumption that a portion of the mass transfer surface is replaced with a new surface by the motion of eddies near the surface, and proposed the following equation for the estimation of liquid-phase mass transfer coefficients:

$$\bar{k} = \sqrt{Ds} \tag{3.25}$$

where s is the rate of surface renewal [s^{-1}]. Equation (3.25) is known as *Danckwerts' penetration model* or the *surface renewal model*. Although the *surface renewal model* offers some insight into the mechanism of mass transfer, no data on the rates of surface renewal are currently available and we cannot apply this model for practical use.

Example 3.1
A 3 mm diameter air bubble is introduced into water from the bottom of a container of depth 0.5 m. Calculate the amount of oxygen absorbed by the single bubble, if it rises at a velocity of 2 m s^{-1}, the pressure inside the bubble is 1 atmosphere (101.325 kPa), and the water temperature is 20 °C. Henry's constant of oxygen in water and the diffusion coefficient of oxygen in water at 20 °C are given as follows:

$$H = 4052 \text{ MPa}, D_L = 2.08 \times 10^{-9} \text{ m}^2 \text{ s}^{-1}.$$

Solution
The equilibrium concentration of oxygen at the surface of the bubble:

$$x_s = (1.01325 \times 10^5)(0.21)/(4.052 \times 10^9) = 5.25 \times 10^{-6} \, [-]$$

Exposure time of the bubble:

$$t_c = (0.5)/(2.0) = 0.25 \text{ s}$$

Molar density of the bulk liquid:

$$c = (1000)/(18.02) = 55.5 \text{ kmol m}^{-3}$$

Mass transfer coefficient according to Eq. (3.24):

$$k_L = (2)\sqrt{(2.08 \times 10^{-9})/(3.14)(0.25)}$$
$$= 1.0 \times 10^{-4} \text{ m s}^{-1}$$

Rate of mass transfer:

$$N_A = (55.5)(1.0 \times 10^{-4})(5.25 \times 10^{-6})$$
$$= 2.31 \times 10^{-8} \text{ kmol m}^{-2} \text{ s}^{-1}$$

Surface area of the bubble:

$$A = (3.14)(3 \times 10^{-3})^2 = 2.86 \times 10^{-5} \text{ m}^2$$

Amount of oxygen absorbed by the single bubble during the rising period:

$$R = (2.31 \times 10^{-8})(2.86 \times 10^{-5})(0.25)$$
$$= 1.65 \times 10^{-10} \, [\text{mol/bubble}]$$

3.5
Physical Interpretation of Some Important Dimensionless Numbers

3.5.1
Reynolds Number

Figure 3.5 shows a flow along a flat plate placed parallel to the flow. Except in the case of an extremely low free stream velocity U_∞ [m s^{-1}], there is a thin layer in which the velocity varies sharply with distance from the wall. This layer is known as the *velocity boundary layer*. Let us consider the order of magnitude of the thickness of the boundary layer δ [m]. From consideration of the order of magnitude of each term of the equation of motion, we have the following relationships:

$$\text{Inertia Force} \approx \rho u \frac{\partial u}{\partial x} \approx \frac{\rho U_\infty^2}{x}$$

$$\text{Viscous Force} \approx \mu \frac{\partial^2 u}{\partial y^2} \approx \mu \frac{U_\infty}{\delta^2}$$

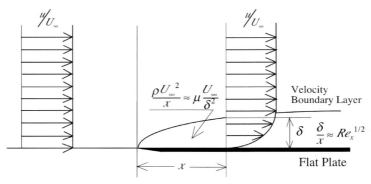

Fig. 3.5 Thickness of velocity boundary layer and Reynolds number.

The equation of motion in the boundary layer can be approximated by the following expression:

$$\frac{\rho U_\infty^2}{x} \approx \mu \frac{U_\infty}{\delta^2} \tag{3.26}$$

where x is the distance from the leading edge of the plate [m], μ is the viscosity [Pa s], and ρ is the density of the fluid [kg m^{-3}].

Rearranging Eq. (3.26), we obtain the following equation for the order of magnitude of the thickness of the boundary layer:

$$\frac{\delta}{x} \approx Re_x^{-1/2} \tag{3.27}$$

where Re_x is the *Reynolds number* defined by the following equation:

$$Re_x \equiv \frac{\rho x U_\infty}{\mu} \quad (3.28)$$

Rearranging Eq. (3.28), we obtain the following equation:

$$Re_x = \frac{\rho U_\infty^2}{(\mu U_\infty / x)} = \frac{\text{inertia force}}{\text{viscous force}} \quad (3.29)$$

This indicates that the Reynolds number is a measure of the thickness of the velocity boundary layer and is also indicative of the relative order of magnitude of the inertia force to the viscous force in the flow of fluid. The transition from laminar to turbulent flow occurs if the Reynolds number exceeds a certain critical value, which is called the *critical Reynolds number*. The critical Reynolds number for a flow inside a circular pipe is about 2300, while that for a flow in a boundary layer along a flat plate is in the range 3.5×10^5 to 1×10^6.

3.5.2
Prandtl Number and Schmidt Number

Figure 3.6 shows the temperature and velocity distributions in a flow along a flat plate. The thin layer near the wall in which the temperature distribution varies sharply is known as the *temperature boundary layer*. If we assume that the thickness of the temperature boundary layer, δ_T [m], is sufficiently small in comparison with that of the velocity boundary layer ($\delta_T \ll \delta$), the velocity at the outer edge of the temperature boundary layer can be approximated by the following equation:

$$u = (\delta_T / \delta) U_\infty \quad (3.30)$$

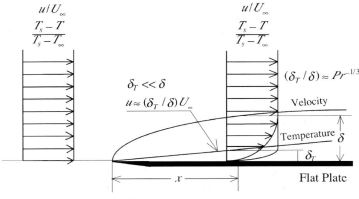

Fig. 3.6 Thickness of thermal boundary layer and Prandtl number.

From consideration of the order of magnitude of each term of the energy equation, we have:

$$\text{Energy Transfer by Convection} \approx \rho c_p u \frac{\partial T}{\partial x} \approx \rho c_p U_\infty \left(\frac{\delta_T}{\delta}\right)\left(\frac{\Delta T}{x}\right)$$

$$\text{Energy Transfer by Conduction} \approx \kappa \frac{\partial^2 T}{\partial y^2} \approx \kappa \frac{\Delta T}{\delta_T^2}$$

The energy equation in the temperature boundary layer can thus be approximated by the following equation:

$$\kappa \frac{(T_s - T_\infty)}{\delta_T^2} \approx \rho c_p U_\infty \left(\frac{\delta_T}{\delta}\right)\left(\frac{T_s - T_\infty}{x}\right) \tag{3.31}$$

where c_p is the specific heat [J kg^{-1} K^{-1}], T_s and T_∞ are the surface and bulk temperatures [K], respectively, x is the distance from the leading edge of the plate [m], κ is the thermal conductivity [W m^{-1} K^{-1}], and ρ is the density of the fluid [kg m^{-3}].

Rearranging Eq. (3.31) by the use of Eq. (3.27), we obtain the following equation:

$$\left(\frac{\delta_T}{\delta}\right) \approx \left(\frac{\kappa}{c_p \mu}\right)^{1/3} = Pr^{-1/3} \tag{3.32}$$

where Pr, the dimensionless number defined by Eq. (3.33), is known as the *Prandtl number*, which represents the relative order of magnitude of the thickness of the temperature boundary layer in comparison with the thickness of the velocity boundary layer.

$$Pr \equiv c_p \mu / \kappa \tag{3.33}$$

For mass transfer in the boundary layer along a flat plate, the concentration distribution also varies sharply near the wall and this thin layer is known as the *concentration boundary layer*. If we assume that the thickness of the concentration boundary layer is sufficiently small in comparison with that of the velocity boundary layer ($\delta_c \ll \delta$), the diffusion equation in the concentration boundary layer is approximated by the following equation in a similar way as in the case of the temperature boundary layer:

$$D \frac{(\omega_s - \omega_\infty)}{\delta_c^2} \approx U_\infty \left(\frac{\delta_c}{\delta}\right)\left(\frac{\omega_s - \omega_\infty}{x}\right) \tag{3.34}$$

Here, D is the diffusion coefficient [m^2 s^{-1}], δ_c is the thickness of the concentration boundary layer [m], and ω_s and ω_∞ are the mass fractions at the surface and in the bulk fluid, respectively.

Rearranging Eq. (3.34) by the use of Eq. (3.27), we obtain the following expression:

$$\left(\frac{\delta_c}{\delta}\right) \approx \left(\frac{\mu}{\rho D}\right)^{-1/3} = Sc^{-1/3} \tag{3.35}$$

where Sc, the dimensionless number defined as in Eq. (3.36), is known as the *Schmidt number*, which represents the relative order of magnitude of the thickness of the concentration boundary layer in comparison with that of the velocity boundary layer.

$$Sc \equiv (\mu/\rho D) \tag{3.36}$$

As is clear from Eqs. (3.33) and (3.36), the Prandtl and Schmidt numbers are dimensionless numbers that depend solely on the physical properties of the systems.

The orders of magnitude of the Prandtl and Schmidt numbers for common gases are as follows:

$$Pr \approx Sc \approx 1$$

and values are almost independent of temperature.

The orders of magnitude of the Prandtl and Schmidt numbers for common liquids, except for liquid metals, are as follows:

$$10 < Pr < 10^2$$

$$400 < Sc < 10^4$$

In contrast to the Prandtl and Schmidt numbers in the gas phase, those in the liquid phase are very much dependent on temperature.

3.5.3
Nusselt Number

The rates of heat transfer in fluid media are usually described in terms of *Nusselt numbers*, as defined by the following equation:

$$Nu \equiv \frac{q}{\kappa(T_s - T_\infty)/L} = \frac{h(T_s - T_\infty)}{\kappa(T_s - T_\infty)/L} = \frac{hL}{\kappa} \tag{3.37}$$

where h is the heat transfer coefficient [W K^{-1} m^{-2}], L is the characteristic length of the heat transfer surface [m], q is the heat flux at the wall [W m^{-2}], and κ is the thermal conductivity of the fluid [W m^{-1} K^{-1}]. Equation (3.37) indicates that the Nusselt number represents the relative order of magnitude of the convective heat flux as compared with the conductive heat flux.

3.5.4
Sherwood Number

The rate of mass transfer is usually described by a dimensionless number, known as the *Sherwood number*. In conventional understanding, Sherwood numbers are taken to be dimensionless numbers composed of the mass transfer coefficient, the diffusivity, and the characteristic length of the system, in analogy to similar numbers in heat transfer, namely the Nusselt numbers. Unfortunately, in practical applications, various definitions of concentration tend to be used, such as mole fraction, partial pressure, partial molar density, mass fraction, and absolute humidity, in a case-by-case way, and hence there are various definitions of mass transfer coefficients and Sherwood numbers. As a result of this variety, we are faced with an unexpected paradox in that numerical values of the Sherwood number obtained from the same mass transfer data but calculated with different mass transfer coefficients do not have to be the same under certain circumstances. For example, the numerical values of (kL/D) and $(k_H L/\rho D)$ are not equal, nor are those of (kL/D) and $(k_L L/D)$. This suggests that there must be something inherently wrong in the conventional definition of the Sherwood number, although we have no such trouble in the case of the Nusselt number.

Here, we propose a new definition of the Sherwood number, which does not depend on the definition of the mass transfer coefficient but on the mass flux or molar flux:

$$\text{Sherwood number} = \frac{\text{mass flux or molar flux}}{\text{characteristic diffusion flux at the interface}} \tag{3.38}$$

Since there are two types of diffusional flux, the mass diffusional flux, $J_A \, (= -\rho D \partial \omega/\partial y)$, and the molar diffusional flux, $J_A^* \, (= -cD\partial x/\partial y)$, the Sherwood numbers defined by Eq. (3.38) are classified into the following two groups:

The mass Sherwood number:

$$Sh = \frac{N_A}{\rho D(\omega_s - \omega_\infty)/L} \tag{3.39}$$

The molar Sherwood number:

$$Sh^* = \frac{(N_A/M_A)}{cD(x_s - x_\infty)/L} \tag{3.40}$$

Here, c is the molar density of the fluid [mol m^{-3}], D is the diffusivity [m^2 s^{-1}], L is the characteristic length of the system [m], M_A is the molecular weight of component *A* [kg kmol^{-1}], N_A is the mass flux of component *A* [kg m^{-2} s^{-1}], x is the mole fraction, ρ is the density of the fluid [kg m^{-3}], and ω is the mass fraction. Table 3.1 summarizes various definitions of mass transfer coefficients and Sherwood numbers.

3.5 Physical Interpretation of Some Important Dimensionless Numbers

Table 3.1 Definitions of mass transfer coefficients and Sherwood numbers.[a]

	Sherwood Number	Mass Transfer Coefficient	
Definition	**Sh or Sh***	**Symbol**	**Definition**
Mass Sherwood Number, Sh, by Eq. (3.39)	$\left(\dfrac{kL}{D}\right)$	k	$N_A = \rho k (\omega_s - \omega_\infty)$
	$\left(\dfrac{k_H L}{\rho D}\right)\left(\dfrac{H_s - H_\infty}{\omega_s - \omega_\infty}\right)$	k_H	$N_A = k_H (H_s - H_\infty)$
Molar Sherwood Number, Sh^*, by Eq. (3.40)	$\left(\dfrac{k_y L}{cD}\right)$	k_y	$N_A^* = k_y (y_s - y_\infty)$
	$\left(\dfrac{k_G RTL}{D}\right)$	k_G	$N_A^* = k_G (p_s - p_\infty)$
	$\left(\dfrac{k_Y L}{cD}\right)\left(\dfrac{Y_s - Y_\infty}{y_s - y_\infty}\right)$	k_Y	$N_A^* = k_Y (Y_s - Y_\infty)$
	$\left(\dfrac{k_L L}{D}\right)$	k_L	$N_A^* = k_L (c_s - c_\infty)$
	$\left(\dfrac{k_x L}{cD}\right)$	k_x	$N_A^* = k_x (x_s - x_\infty)$
	$\left(\dfrac{k_X L}{cD}\right)\left(\dfrac{X_s - X_\infty}{x_s - x_\infty}\right)$	k_X	$N_A^* = k_X (X_s - X_\infty)$

a) c = molar density [kmol m^{-3}], D = diffusivity [m^2 s^{-1}], H = absolute humidity [–], L = characteristic length [m], M_A = molecular weight [kg kmol^{-1}], N_A = mass flux [kg m^{-2} s^{-1}], $N_A^* = N_A/M_A$ = molar flux [kmol m^{-2} s^{-1}], p = partial pressure [Pa], R = gas constant [J mol^{-1} K^{-1}], T = temperature [K], x = mole fraction of liquid [–], y = mole fraction of gas [–], $X = x/(1-x)$, $Y = y/(1-y)$, ρ = density [kg m^{-3}], ω = mass fraction [–]

The fact that the values of the two Sherwood numbers, Sh and Sh^*, are unequal even for the same mass transfer data may seem quite strange. From the definitions of the Sherwood numbers, Eqs. (3.39) and (3.40), we have the following equation:

$$\frac{Sh^*}{Sh} = \left\{1 + \left(\frac{M_B}{M_A} - 1\right)\omega_\infty\right\} \qquad (3.41)$$

Equation (3.41) indicates that the ratio of the two Sherwood numbers, Sh^*/Sh, usually depends on the molecular weight ratio and on the bulk concentrations, as shown in Fig. 3.7.

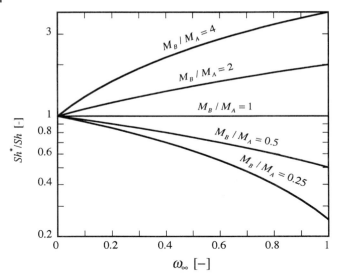

Fig. 3.7 Mass and molar Sherwood numbers: effect of molecular weight and bulk concentration.

3.5.5
Dimensionless Numbers Commonly Used in Heat and Mass Transfer

Table 3.2 summarizes some important dimensionless numbers commonly used in heat and mass transfer problems.

Example 3.2
Calculate the mass transfer coefficients k, k_H, k_y and the Sherwood numbers under the following conditions.

$$N_A = 1.0 \times 10^{-4} \text{ kg m}^{-2} \text{ s}^{-1},\ M_A = 18 \text{ kg kmol}^{-1},\ M_B = 29 \text{ kg kmol}^{-1},$$
$$D_G = 1.0 \times 10^{-5} \text{ m}^2 \text{ s}^{-1},\ \rho_G = 1.0 \text{ kg m}^{-3},\ \omega_s = 0.6,\ \omega_\infty = 0.1$$

Solution
Step 1) Conversion of concentration:

$$y_s = \frac{(0.6)/(18)}{(0.6)/(18) + (1 - 0.6)/(29)} = 0.707$$

$$y_\infty = \frac{(0.1)/(18)}{(0.1)/(18) + (1 - 0.1)/(29)} = 0.152$$

Table 3.2 Some important dimensionless numbers used in heat and mass transfer.

	Symbol	Name	Definition	Remark
Fluxes	Nu	Nusselt number	$\dfrac{hL}{\kappa}$	Rates of heat transfer
	Sh	Sherwood number	$\dfrac{N_A L}{\rho D (\omega_s - \omega_\infty)}$	Rates of mass transfer
	St_H	Stanton number for heat transfer	$\dfrac{Nu}{RePr}$	Rate of turbulent heat transfer
	St_M	Stanton number for mass transfer	$\dfrac{Sh}{ReSc}$	Rate of turbulent mass transfer
	j_D	j-factor for mass transfer	$St_M Sc^{2/3}$	Rate of turbulent mass transfer
	j_H	j-factor for heat transfer	$St_H Pr^{2/3}$	Rate of turbulent heat transfer
Operating conditions and physical properties	Fo_H	Fourier number for heat transfer	$\dfrac{\kappa t}{\rho c_p L^2}$	Transient heat transfer
	Fo_M	Fourier number for mass transfer	$\dfrac{Dt}{L^2}$	Transient mass transfer
	Gz_H	Graetz number for heat transfer	$\left(\dfrac{\pi D_T}{4L}\right) RePr$	Heat transfer inside a circular pipe
	Gz_M	Graetz number for mass transfer	$\left(\dfrac{\pi D_T}{4L}\right) ReSc$	Mass transfer inside a circular pipe
	Gr	Grashof number	$\dfrac{g\beta \Delta T^3 L^2 \rho^2}{\mu^2}$	Natural convection
	Pe_H	Peclet number for heat transfer	$RePr$	Ratio of heat conduction to convective heat flux
	Pe_M	Peclet number for mass transfer	$ReSc$	Ratio of diffusion to convective mass flux
	Pr	Prandtl number	$\dfrac{c_p \mu}{\kappa}$	Thickness of temperature boundary layer
	Re	Reynolds number	$\dfrac{\rho L U_\infty}{\mu}$	Thickness of velocity boundary layer
	Sc	Schmidt number	$\dfrac{\mu}{\rho D}$	Thickness of concentration boundary layer
	B	Transfer number for mass transfer	$\left(\dfrac{\omega_s - \omega_\infty}{1 - \omega_s}\right)$	Dimensionless concentration driving force

$$H_s = 0.6/(1-0.6) = 1.5$$

$$H_\infty = 0.1/(1-0.1) = 0.111$$

$$\overline{M_s} = (18)(0.707) + (29)(1-0.707) = 21.2 \text{ kg kmol}^{-1}$$

$$c_s = (1.0)/(21.2) = 0.0472 \text{ kmol m}^{-3}$$

Step 2) Calculation of mass transfer coefficients:

$$k = \frac{(1.0 \times 10^{-4})}{(1.0)(0.6 - 0.1)} = 2.0 \times 10^{-4} \text{ m s}^{-1}$$

$$k_y = \frac{(1.0 \times 10^{-4})/(18)}{(0.707 - 0.152)} = 1.00 \times 10^{-5} \text{ kmol m}^{-2} \text{ s}^{-1}$$

$$k_H = \frac{(1.0 \times 10^{-4})}{(1.5 - 0.111)} = 0.720 \times 10^{-4} \text{ kg m}^2 \text{ s}^{-1}$$

Step 3) Calculation of mass Sherwood number Sh by reference to Tab. 3.1:

$$Sh = \frac{kL}{D_G} = \frac{(2.0 \times 10^{-4})(0.1)}{(1.0 \times 10^{-5})} = 2.00$$

$$Sh = \left(\frac{k_H L}{\rho D}\right)\left(\frac{H_s - H_\infty}{\omega_s - \omega_\infty}\right)$$

$$= \frac{(0.720 \times 10^{-4})(0.1)}{(1.0)(1.0 \times 10^{-5})} \left(\frac{1.5 - 0.111}{0.6 - 0.1}\right)$$

$$= (0.720)(2.778) = 2.00$$

However, when we calculate the numerical value of the dimensionless term $(k_H L/\rho_s D_G)$, we obtain a quite different value from that of the true Sherwood number ($Sh = 2.0$), as shown below:

$$\frac{k_H L}{\rho_s D_G} = \frac{(0.720 \times 10^{-4})(0.1)}{(1.0)(1.0 \times 10^{-5})} = 0.720$$

Step 4) Calculation of the molar Sherwood number:

$$Sh^* = \frac{k_y L}{c_s D_G} = \frac{(1.00 \times 10^{-5})(0.1)}{(0.0472)(1.0 \times 10^{-5})} = 2.12$$

$$\frac{Sh^*}{Sh} = \frac{2.12}{2.00} = 1.06$$

The numerical value of the molar Sherwood number, Sh^*, is 1.06 times larger than the mass Sherwood number, Sh. If we calculate the ratio of the two Sherwood numbers, Sh^*/Sh, using Eq. (3.41), the same result is obtained:

$$\left(\frac{Sh^*}{Sh}\right) = \left\{1 + \left(\frac{29}{18} - 1\right)(0.1)\right\} = 1.06$$

3.6 Dimensional Analysis

3.6.1 Principle of Similitude and Dimensional Homogeneity

The behavior of the physical processes in actual problems is affected by so many physical quantities of various dimensions that a complete mathematical description thereof is usually very difficult and sometimes practically impossible due to the complicated nature of the phenomena. We know from experience that if two systems are geometrically similar there usually exists some kind of similarity under certain conditions, such as kinematic similarity, dynamic similarity, thermal similarity, and similarity of concentration distribution, and that if similarity conditions are satisfied we can greatly reduce the number of independent variables required to describe the behavior of the process. In this way, we can systematically understand, describe, and even predict the behavior of physical processes in real problems in a relatively simple manner. This principle is known as the *principle of similitude*. *Dimensional analysis* is a method of deducing logical groupings of the variables, through which we can describe similarity criteria of the processes.

Physical quantities such as length [L], mass [M], time [T], and temperature [Θ] are dimensional quantities and the magnitude of each quantity can be described by multiples of the unit of each dimension, namely m, kg, s, and K, respectively. Through experience, we can select a certain number of *fundamental dimensions*, such as those mentioned above, and express all other dimensional quantities in terms of products of powers of these fundamental dimensions. Furthermore, in describing the behavior of physical processes, we know that there is an implicit principle that we cannot add or subtract physical quantities of different dimensions. This means that the equations governing physical processes must be dimensionally consistent and each term of the equation must have the same dimensions. This principle is known as the *principle of dimensional homogeneity*.

Let us investigate, for instance, the dimensions of each term of the equation of motion for a one-dimensional steady flow in a horizontal circular pipe.

3 Governing Equations of Mass Transfer

$$\rho u \frac{\partial u}{\partial z} = \frac{\partial p}{\partial z} + \mu \frac{\partial^2 u}{\partial r^2} \tag{3.42}$$

The dimensions of each term of Eq. (3.42) are as follows:
Left-hand side:

$$\left[\frac{M}{L^3}\right]\left[\frac{L}{T}\right]\left[\frac{L}{T}\right]\left[\frac{1}{L}\right] = \left[\frac{M}{L^2 T^2}\right]$$

First term on the right-hand side:

$$\left[\frac{F}{L^2}\right]\left[\frac{1}{L}\right] = \left[\frac{M}{LT^2}\right]\left[\frac{1}{L}\right] = \left[\frac{M}{L^2 T^2}\right]$$

The second term on the right-hand side:

$$\left[\frac{M}{LT}\right]\left[\frac{L}{T}\right]\left[\frac{1}{L^2}\right] = \left[\frac{M}{L^2 T^2}\right]$$

This clearly indicates that the above theoretical equation automatically satisfies the condition of *the principle of dimensional homogeneity* and we say that the equation is dimensionally perfect. From dimensionally perfect equations we can obtain a combination of dimensionless numbers to describe the similarity criteria of the physical process.

3.6.2
Finding Dimensionless Numbers and Pi Theorem

Rearrangement of theoretical equations: If a theoretical equation governing a process is known and is dimensionally perfect, we can obtain a complete set of dimensionless numbers by rearrangement of the equation in a suitable way. For example, Eq. (3.42) can be rewritten in dimensionless form by introducing the following dimensionless variables:

Dimensionless velocity: $u' = u/U_{av}$
Dimensionless pressure: $p' = p/\rho U_{av}^2$
Dimensionless axial distance: $z' = z/d$
Dimensionless radial distance: $r' = r/d$

Substituting the above variables into Eq. (3.42), we obtain the following equation:

$$u' \frac{\partial u'}{\partial z'} = \frac{\partial p'}{\partial z'} + \frac{1}{Re} \frac{\partial^2 u'}{\partial r'^2} \tag{3.43}$$

This indicates that the relationship between dimensionless pressure and dimensionless velocity will be similar if Reynolds numbers in two flows, $Re \, (\equiv \rho U_{av} d / \mu)$, are equal. Indeed, the criterion for similarity in two flows is equality of the Reynolds numbers.

Rayleigh's method of indices: Although rearrangement of the theoretical equations provides us with complete sets of the dimensionless numbers that are required to describe the similarity criteria for a process, we are sometimes confronted with much more difficult cases in which we cannot formulate any mathematical equation for the process due to its overwhelming complexity. Under these circumstances, *Rayleigh's method of indices*, as described in the following, may offer some help in obtaining the necessary dimensionless numbers to describe the similarity criteria of a process, though not perfectly as in the case of the rearrangement of the theoretical equations, but more approximately, which suggests a possible combination of some dimensionless numbers.

Let us consider, for example, the pressure drop in a circular pipe. We will assume that the pressure drop Δp [N m^{-2}] is a function of the viscosity μ [Pa s], the density ρ [kg m^{-3}], the velocity u [m s^{-1}], and the length l [m] and diameter d [m] of the pipe:

$$\Delta p = f(\mu, \rho, u, l, d) \tag{3.44}$$

If we further assume that the mathematical equation can be written in terms of products of powers of the above mentioned variables, we have the following equation:

$$\Delta p = C \mu^a \rho^b u^c l^d d^e \tag{3.45}$$

Rewriting the above equation in terms of the fundamental dimensions, we obtain the following equation:

$$\left[\frac{M}{LT^2} \right] = \left[\frac{M}{LT} \right]^a \left[\frac{M}{L^3} \right]^b \left[\frac{L}{T} \right]^c [L]^d [L]^e \tag{3.46}$$

from which we derive the following simultaneous equations:

dimension of the mass [M]: $1 = a + b$ (3.47a)
dimension of the length [L]: $-1 = -a - 3b + c + d + e$ (3.47b)
dimension of the time [T]: $-2 = -a - c$ (3.47c)

The solutions of Eq. (3.47) in terms of b, c, and e are as follows:

$b = 1 - a$ (3.48a)
$c = 2 - a$ (3.48b)
$e = a - d$ (3.48c)

Substituting the above equations into Eq. (3.45), we obtain the following equation:

$$\left(\frac{\Delta p}{\rho u^2}\right) = C \left(\frac{\rho du}{\mu}\right)^{-a} \left(\frac{l}{d}\right)^d \tag{3.49}$$

which indicates that the dimensionless pressure drop, $(\Delta p/\rho u^2)$, is a function of two independent variables, namely the dimensionless pipe length, (l/d), and the Reynolds number, $(Re (= \rho du/\mu))$.

On the other hand, if we choose another combination of independent variables, such as in terms of a, c, and e, we have:

$$a = 1 - b \tag{3.50a}$$
$$c = 1 + b \tag{3.50b}$$
$$e = 1 + b - d \tag{3.50c}$$

Substituting these values into Eq. (3.45), we obtain the following equation:

$$\left(\frac{\Delta pd}{\mu u}\right) = \left(\frac{\Delta p}{\rho u^2}\right)\left(\frac{\rho du}{\mu}\right) = C\left(\frac{\rho du}{\mu}\right)^b \left(\frac{l}{d}\right)^d \tag{3.51}$$

The dimensionless term on the left-hand side of Eq. (3.51), $(\Delta pd/\mu u)$, is the product of two dimensionless numbers, namely the dimensionless pressure drop, $(\Delta p/\rho u^2)$ and the Reynolds number, $(\rho ud/\mu)$. This indicates that a new dimensionless number can be defined as the product of the powers of some other dimensionless numbers. Here we have a dilemma, in that we have to decide which combination of the dimensionless numbers, that is, by Eq. (3.49) or (3.51), is more suitable for describing the actual physical phenomenon at hand. *Rayleigh's method of indices* does not give any answer to this question. An important fact about *Rayleigh's method* is that *it only suggests a possible combination of dimensionless numbers*, which needs experimental verification. Moreover, the initial selection of dimensional variables will greatly affect the final results, and obtaining a suitable combination of the dimensionless numbers is a matter of experience and know-how. Nevertheless, it must be stressed that in an experimental approach *Rayleigh's method of indices* is a powerful tool.

Buckingham's Pi Theorem: Buckingham [2] showed that an important theorem links n_π independent dimensionless numbers, n_1 physical quantities of importance, and n_2 fundamental dimensions, which is known as the *pi theorem* and can be written by the following equation:

$$n_\pi = n_1 - n_2 \tag{3.52}$$

For example, if we apply *Buckingham's pi theorem* to the above-mentioned example, we have the following equations:

Number of physical quantities = 6
Number of fundamental dimensions = 3
Number of independent dimensionless numbers = 6 − 3 = 3

The same conclusion is obtained from the results of Rayleigh's method of indices.

References

1 R. B. Bird, W. E. Stewart, and E. N. Lightfoot, "*Transport Phenomena*", pp. 83–91, 318–319, 559, Wiley (1960).
2 E. Buckingham, "Model Experiments and the Forms of Empirical Equations", *Transactions of the American Society of Mechanical Engineers*, **35**, 263–296 (1915).
3 H. S. Carslaw and J. C. Jaeger, "*Conduction of Heat in Solids*", 2nd Edition, pp. 58–62, Oxford Univ. Press (1959).
4 P. V. Danckwerts, "Significance of Liquid-Film Coefficients in Gas Absorption", *Industrial and Engineering Chemistry*, **43**, 1460–1467 (1951).
5 R. Higbie, "The Rate of Absorption of a Pure Gas into a Still Liquid During Short Periods of Exposure", *Transactions of the American Institute of Chemical Engineers*, **36**, 365–389 (1935).
6 W. K. Lewis and W. Whitman, "Principles of Gas Absorption", *Industrial and Engineering Chemistry*, **16**, [12], 1215–1220 (1924).
7 Lord Rayleigh, "The Principle of Similitude", *Nature*, **95**, 66–68 (1915).
8 O. Reynolds, "An Experimental Investigation of the Circumstances which determine whether the Motion of Water shall be Direct or Sinuous, and of the Law of Resistances in Parallel Channels", *Philosophical Transactions*, **174**, 935–982 (1883).
9 H. Schlichting, "*Boundary Layer Theory*", 6th Edition, pp. 44–62, 252–255, McGraw-Hill (1968).

4
Mass Transfer in a Laminar Boundary Layer

4.1
Velocity Boundary Layer

Heat or mass transfer in a laminar flow can be described by exact solutions of the governing equations, such as the continuity equation, the equation of motion, the energy equation, and the diffusion equation, under the given boundary conditions. Since the energy and diffusion equations contain velocity terms, we first have to solve the equation of motion to obtain flow fields for the given problems before we can deal with heat or mass transfer problems. Because of the nonlinear nature of the Navier–Stokes equation, obtaining the exact solution is mathematically difficult even for the simplest problem of laminar flow along a flat plate. In this chapter, we describe some fundamental aspects of mass transfer in a laminar boundary layer along a flat plate.

4.1.1
Boundary Layer Equation

We consider a two-dimensional flow of a fluid of constant physical properties along a flat plate at zero incidence as shown in Fig. 4.1. In the region near the wall, the velocity of the fluid is much decelerated due to the effect of its viscosity, but in the region far away from the wall the effect becomes less significant and the velocity gradually approaches the free stream velocity as the distance from the wall increases. The region near the wall, in which the effect of viscosity is considerable, known as the boundary layer, becomes smaller and smaller as the free stream velocity increases. Except at very low free stream velocity, the flow is divided into two regions, the boundary layer in which the effect of viscosity is considerable, and the flow of the ideal fluid.

L. Prandtl was the first to derive the so-called boundary layer equations for the flow near the wall where the effect of viscosity is considerable, by considering the order of magnitude of each term in the Navier–Stokes equation [12].

If we tentatively assume that the orders of magnitude of u and x are both unity in Eq. (3.13.a) and that the thickness of the boundary layer is δ ($\delta \ll 1$), then from the continuity equation v and y will be of the same order of magni-

Mass Transfer. From Fundamentals to Modern Industrial Applications. Koichi Asano
Copyright © 2006 WILEY-VCH Verlag GmbH & Co. KGaA, Weinheim
ISBN: 3-527-31460-1

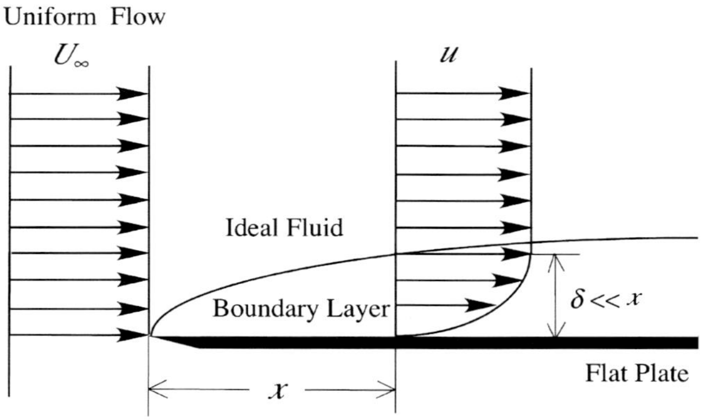

4 Mass Transfer in a Laminar Boundary Layer

Uniform Flow

Fig. 4.1 Boundary layer along a flat plate at zero incidence.

tude as δ. Therefore, the order of magnitude of each term of Eq. (3.13.a) can be written as follows:

$$u\frac{\partial u}{\partial x} + v\frac{\partial u}{\partial y} = -\frac{1}{\rho}\frac{\partial P}{\partial x} + \left(\frac{\mu}{\rho}\right)\left(\frac{\partial^2 u}{\partial x^2} + \frac{\partial^2 u}{\partial y^2}\right) \tag{3.13a}$$

$$(1\cdot 1) \quad \left(\delta\cdot\frac{1}{\delta}\right) \quad (1) \quad \delta^2\left(1 \quad \frac{1}{\delta^2}\right)$$

From the condition that the viscous term and the inertia term should be of the same order of magnitude in the boundary layer, the second term on the right-hand side of the above equation must contain δ^2. If we eliminate the higher order terms of δ in the equation, Eq. (3.13.a) can be approximated by the following equation:

$$u\frac{\partial u}{\partial x} + v\frac{\partial u}{\partial y} = -\frac{1}{\rho}\frac{\partial P}{\partial x} + \frac{\mu}{\rho}\frac{\partial^2 u}{\partial y^2} \tag{4.1}$$

Similarly, the order of magnitude of each term in Eq. (3.13.b) can be written as:

$$u\frac{\partial v}{\partial x} + v\frac{\partial v}{\partial y} = -\frac{1}{\rho}\frac{\partial P}{\partial y} + \left(\frac{\mu}{\rho}\right)\left(\frac{\partial^2 v}{\partial x^2} + \frac{\partial^2 v}{\partial y^2}\right) \tag{3.13b}$$

$$(1\cdot\delta) \quad (\delta\cdot 1) \quad 1 \quad \delta^2\left(\delta \quad \frac{\delta}{\delta^2}\right)$$

This indicates that the order of magnitude of the first term on the right-hand side of Eq. (3.13.b) is the order of δ:

$$\frac{\partial P}{\partial y} = 0 \tag{4.2}$$

Eq. (4.2) indicates that the pressure within the boundary layer is equal to that in the free stream, and so Bernoulli's equation for ideal fluids is applicable to estimate the pressure within the boundary layer.

$$P + \frac{1}{2}\rho U^2 = \text{constant} \tag{4.3}$$

Rearranging Eq. (4.1) by the use of Eq. (4.3), we obtain the following equation:

$$u\frac{\partial u}{\partial x} + v\frac{\partial u}{\partial y} = U\frac{\partial U}{\partial x} + \frac{\mu}{\rho}\frac{\partial^2 u}{\partial y^2} \tag{4.4}$$

where U is the free stream velocity [m s^{-1}].

In a similar way, if the effects of radiation and viscous dissipation are negligibly small, the energy equation for the laminar boundary layer along a flat plate can be written as follows:

$$u\frac{\partial T}{\partial x} + v\frac{\partial T}{\partial y} = \left(\frac{\kappa}{\rho c_p}\right)\frac{\partial^2 T}{\partial y^2} \tag{4.5}$$

The diffusion equation for binary fluid mixtures with constant physical properties and without chemical reaction in the laminar boundary layer can be written as follows:

$$u\frac{\partial \omega}{\partial x} + v\frac{\partial \omega}{\partial y} = D\frac{\partial^2 \omega}{\partial y^2} \tag{4.6}$$

4.1.2
Similarity Transformation

For flow along a flat plate at zero incidence, the condition of $dU/dx = 0$ holds and the continuity equation and the equation of motion for fluids of constant physical properties can be written as:

$$\frac{\partial u}{\partial x} + \frac{\partial v}{\partial y} = 0 \tag{4.7}$$

$$u\frac{\partial u}{\partial x} + v\frac{\partial u}{\partial y} = \frac{\mu}{\rho}\frac{\partial^2 u}{\partial y^2} \tag{4.8}$$

The boundary conditions are:

$$y = 0: \quad u = v = 0$$
$$y = \delta: \quad u = U_\infty$$

If we assume that the velocity distribution in the boundary layer is similar, the dimensionless velocity, u/U_∞, can be expressed as a function of the dimensionless distance from the wall, y/δ, where δ is the thickness of the boundary layer. Considering the equation for the thickness of the boundary layer, Eq. (3.27), the dimensionless distance from the wall, η, can be defined by the following equation:

$$\eta \equiv y \left(\frac{\rho U_\infty}{\mu x} \right)^{1/2} \approx \left(\frac{y}{\delta} \right) \tag{4.9}$$

The stream function, $\psi(x,y)$, which automatically satisfies the continuity equation, can easily be converted into dimensionless form:

$$u \equiv \frac{\partial \psi}{\partial y} \tag{4.10a}$$

$$v \equiv -\frac{\partial \psi}{\partial x} \tag{4.10b}$$

$$\psi = \left(\frac{\mu x U_\infty}{\rho} \right)^{1/2} F(\eta) \tag{4.11}$$

where $F(\eta)$ is the dimensionless stream function.

The x- and y-components of the velocity in the boundary layer, u and v in Eq. (4.10), can be written by the following equations:

$$\frac{u}{U_\infty} = \frac{d}{d\eta} F(\eta) \tag{4.12a}$$

$$\frac{v}{U_\infty} = \frac{1}{2} \left(\frac{1}{Re_x} \right)^{1/2} \left(\eta \frac{d}{d\eta} F(\eta) - F(\eta) \right) \tag{4.12b}$$

Substituting Eqs. (4.12.a) and (4.12.b) into Eq. (4.8), we obtain the boundary layer equation in terms of the dimensionless stream function as:

$$\frac{d^3 F}{d\eta^3} + \frac{1}{2} F \frac{d^2 F}{d\eta^2} = 0 \tag{4.13}$$

The boundary conditions are as follows:

$$\eta = 0 : F = \frac{dF}{d\eta} = 0$$

$$\eta = \infty : \frac{dF}{d\eta} = 1$$

4.1.3
Integral Form of the Boundary Layer Equation

The boundary layer equation, Eq. (4.13), is nonlinear in nature. As a result, obtaining an analytical solution of the equation is impossible and only a numerical approach is possible.

Rearranging Eq. (4.13), we have:

$$\frac{d}{d\eta}\left(\frac{d^2 F}{d\eta^2}\right) + \frac{F}{2}\left(\frac{d^2 F}{d\eta^2}\right) = 0 \tag{4.14}$$

Integrating Eq. (4.14) by separation of the variable, we have:

$$\ln\left\{\frac{d}{d\eta}\left(\frac{dF}{d\eta}\right)\right\} = \int_0^\eta \left(\frac{-F}{2}\right) d\eta \tag{4.15}$$

Integrating Eq. (4.15) by the use of boundary conditions, we obtain the following equation:

$$\frac{u}{U_\infty} = F' = \frac{\int_0^\eta \exp\left(\int_0^\eta \left(\frac{-F}{2}\right) d\eta\right) d\eta}{\int_0^\infty \exp\left(\int_0^\eta \left(\frac{-F}{2}\right) d\eta\right) d\eta} \tag{4.16}$$

L. Howarth [5] carried out numerical integration of the boundary layer equation and tabulated numerical values of $F(\eta)$ and their derivatives. Figure 4.2 shows the velocity distribution in a boundary layer along a flat plate at zero incidence. Table 4.1 shows the numerical solutions after Howarth.

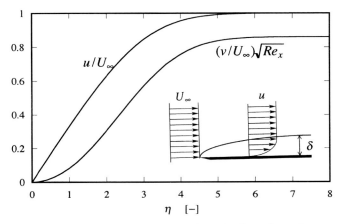

Fig. 4.2 Velocity distributions in a laminar boundary layer along a flat plate at zero incidence after L. Howarth [5].

Table 4.1 Numerical solutions of the stream function for a laminar boundary layer along a flat plate at zero incidence after L. Howarth [5].

η	F	F'	F''	η	F	F'	F''
0	0.00000	0.00000	0.33206	4.6	2.88826	0.98269	0.02948
0.2	0.00664	0.06641	0.33199	4.8	3.08534	0.98779	0.02187
0.4	0.02656	0.13277	0.33147	5.0	3.28329	0.99155	0.01591
0.6	0.05974	0.19894	0.33008	5.2	3.48189	0.99425	0.01134
0.8	0.10611	0.26471	0.32739	5.4	3.68094	0.99616	0.00793
1.0	0.16557	0.32979	0.32301	5.6	3.88031	0.99748	0.00543
1.2	0.23795	0.39378	0.31659	5.8	4.07990	0.99838	0.00365
1.4	0.32298	0.45627	0.30787	6.0	4.27964	0.99898	0.00240
1.6	0.42032	0.51676	0.29667	6.2	4.47948	0.99937	0.00155
1.8	0.52952	0.57477	0.28293	6.4	4.67938	0.99961	0.00098
2.0	0.65003	0.62977	0.26675	6.6	4.87931	0.99977	0.00061
2.2	0.78120	0.68132	0.24835	6.8	5.07928	0.99987	0.00037
2.4	0.92230	0.72899	0.22809	7.0	5.27926	0.99992	0.00022
2.6	1.07252	0.77246	0.20646	7.2	5.47925	0.99996	0.00013
2.8	1.23099	0.81152	0.18401	7.4	5.67924	0.99998	0.00007
3.0	1.39682	0.84605	0.16136	7.6	5.87924	0.99999	0.00004
3.2	1.56911	0.87609	0.13913	7.8	6.07923	1.00000	0.00002
3.4	1.74696	0.90177	0.11788	8.0	6.27923	1.00000	0.00001
3.6	1.92954	0.92333	0.09809	8.2	6.47923	1.00000	0.00001
3.8	2.11605	0.94112	0.08013	8.4	6.67923	1.00000	0.00000
4.0	2.30576	0.95552	0.06424	8.6	6.87923	1.00000	0.00000
4.2	2.49806	0.96696	0.05052	8.8	7.07923	1.00000	0.00000
4.4	2.69238	0.97587	0.03897				

4.1.4
Friction Factor

The viscous drag due to friction in a boundary layer along a flat plate can easily be obtained from the dimensionless stream function. The wall shear stress at x [m] from the leading edge, τ_{wall} [N m^{-2}], is obtained by applying Newton's law of viscosity:

$$\tau_{wall} = \mu \frac{\partial u}{\partial y}\bigg|_{y=0} = \mu U_\infty \left(\frac{\rho U_\infty}{\mu x}\right)^{1/2} F''(0) \qquad (4.17)$$

If we substitute a numerical value of $F''(0) = 0.332$ into the above equation, the local friction factor at x [m] from the leading edge, f_x [–], can be written as:

$$f_x \equiv \frac{\tau_{wall}}{(\rho U_\infty^2/2)} = \frac{2F''(0)}{Re_x^{1/2}} = 0.664\, Re_x^{-1/2} \qquad (4.18)$$

where Re_x is the *local Reynolds number* defined by the following equation:

$$Re_x \equiv \rho U_\infty x/\mu \tag{4.19}$$

The average friction factor, $\bar{f}$, over the entire length of the plate, L [m], can be written by the following equation:

$$\bar{f} \equiv \frac{\overline{\tau_w}}{(\rho U_\infty^2/2)} = \frac{1}{L}\int_0^L \frac{0.664}{Re_x^{1/2}}\,dx = 1.328\, Re^{-1/2} \tag{4.20}$$

where Re is the *average Reynolds number* defined by the following equation:

$$Re \equiv \rho U_\infty L/\mu \tag{4.21}$$

4.2
Temperature and Concentration Boundary Layers

4.2.1
Temperature and Concentration Boundary Layer Equations

Figure 4.3 shows a laminar boundary layer along a flat plate at zero incidence, in which the velocity and the concentration (or temperature) boundary layers develop simultaneously. If the surface temperature and the free stream temperature are T_s [K] and T_∞ [K], respectively, or the surface (interface) concentration and the free stream concentration are ω_s [–] and ω_∞ [–], the dimensionless temperature and the dimensionless concentration can be defined as:

$$\theta_T = \frac{T_s - T}{T_s - T_\infty} \tag{4.22a}$$

$$\theta_c = \frac{\omega_s - \omega}{\omega_s - \omega_\infty} \tag{4.22b}$$

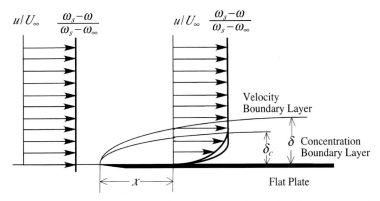

Fig. 4.3 Velocity and concentration boundary layers along a flat plate at zero incidence.

Substituting Eqs. (4.22.a) and (4.22.b) and Eqs. (4.12.a) and (4.12.b) into the energy equation, Eq. (4.5), and the diffusion equation, Eq. (4.6), we obtain the following equations in terms of the dimensionless stream functions and the temperature or concentration:

$$\frac{d^2\theta_T}{d\eta^2} + \frac{Pr}{2} F \frac{d\theta_T}{d\eta} = 0 \tag{4.23}$$

$$\frac{d^2\theta_c}{d\eta^2} + \frac{Sc}{2} F \frac{d\theta_c}{d\eta} = 0 \tag{4.24}$$

The boundary conditions are:

$\eta = 0: \quad \theta_T = \theta_c = 0$
$\eta = \infty: \quad \theta_T = \theta_c = 1$

4.2.2
Integral Form of Thermal and Concentration Boundary Layer Equations

Integrating Eq. (4.23) in a similar way as in the case of the velocity boundary layer equation, Eq. (4.16), we obtain the following equation for the dimensionless temperature distribution:

$$\theta_T = \frac{\int_0^\eta \exp\left(\int_0^\eta \left(\frac{-Pr}{2}F\right) d\eta\right) d\eta}{\int_0^\infty \exp\left(\int_0^\eta \left(\frac{-Pr}{2}F\right) d\eta\right) d\eta} \tag{4.25}$$

Comparison of Eq. (4.25) with Eq. (4.16) indicates that the distribution of the dimensionless temperatures coincides with that of the dimensionless velocities if the Prandtl number is unity ($Pr = 1$).

From Eq. (4.25), the dimensionless temperature gradient is given by the following equation:

$$\theta_T'(0) = \frac{1}{\int_0^\infty \exp\left(\int_0^\eta \left(\frac{-Pr}{2}F\right) d\eta\right) d\eta} \tag{4.26}$$

The dimensionless local heat flux at a distance x [m] from the leading edge of the plate, or the local Nusselt number, can easily be obtained from Fourier's law of heat conduction:

$$Nu_x \equiv \frac{q_x}{\{\kappa(T_s - T_\infty)/x\}} = Re_x^{1/2} \theta_T'(0) \tag{4.27}$$

4.2 Temperature and Concentration Boundary Layers

Similar equations are obtained for the dimensionless concentration distributions:

$$\theta_c = \frac{\int_0^\eta \exp\left(\int_0^\eta \left(\frac{-Sc}{2} F\right) d\eta\right) d\eta}{\int_0^\infty \exp\left(\int_0^\eta \left(\frac{-Sc}{2} F\right) d\eta\right) d\eta} \quad (4.28)$$

$$\theta_c'(0) = \frac{1}{\int_0^\infty \exp\left(\int_0^\eta \left(\frac{-Sc}{2} F\right) d\eta\right) d\eta} \quad (4.29)$$

Again, the distribution of the dimensionless concentration coincides with that of the velocity if the Schmidt number is unity ($Sc = 1$).

For unidirectional diffusion, as described in Section 2.2.2, the dimensionless local rate of mass transfer at a distance x [m], or the local Sherwood number, can be obtained from the following equation:

$$Sh_x (1 - \omega_s) \equiv \frac{N_{Ax}(1-\omega_s)}{\{\rho D (\omega_s - \omega_\infty)/x\}} = Re_x^{1/2} \theta_c'(0) \quad (4.30)$$

Example 4.1
Calculate the concentration distribution and the concentration gradient at the wall for a laminar boundary layer along a flat plate by use of the numerical solution after Howarth shown in Tab. 4.1 if the Schmidt number is two ($Sc = 2.0$).

Solution
Substituting the numerical value of $Sc = 2$ into Eq. (4.28), we obtain the following equations:

$$I_1(\eta) \equiv \int_0^\eta F(\eta) \, d\eta = \sum_0^j \tfrac{1}{2}(F_j + F_{j+1}) \Delta \eta \quad \text{(A-1)}$$

$$I_2(\eta) \equiv \exp\{-I_1(\eta)\} \quad \text{(A-2)}$$

$$I_3(\eta) \equiv \int_0^\eta I_2(\eta) \, d\eta = \sum_0^{j+1} \tfrac{1}{2}(I_{2,j} + I_{2,j+1}) \Delta \eta \quad \text{(A-3)}$$

$$\theta_c = I_3(\eta)/I_3(\infty) \quad \text{(A-4)}$$

From Eq. (4.29), we have:

$$\theta_c'(0) = 1/I_3(\infty) \quad \text{(A-5)}$$

Table 4.2 shows the results of the numerical solution. The numerical value in the last row of the fifth column of the table represents $I_3(\infty)$. Substituting this va-

4 Mass Transfer in a Laminar Boundary Layer

Table 4.2 Results of the calculation in Example 4.1.

η	$F(\eta)$	$I_1(\eta)$	$I_2(\eta)$	$I_3(\eta)$	$\theta(\eta)$
0.0	0.00000	0.00000	1.00000	0.00000	0.00000
0.4	0.02560	0.00512	0.99489	0.39898	0.16937
0.8	0.10611	0.03146	0.96903	0.79176	0.33611
1.2	0.23795	0.10027	0.90459	1.16649	0.49519
1.6	0.42032	0.23193	0.79300	1.50600	0.63931
2.0	0.65003	0.44600	0.64019	1.79264	0.76099
2.4	0.92230	0.76046	0.46745	2.01417	0.85504
2.8	1.23099	1.19112	0.30388	2.16844	0.92052
3.2	1.56911	1.75114	0.17358	2.26393	0.96106
3.6	1.92954	2.45087	0.08622	2.31589	0.98312
4.0	2.30576	3.29793	0.03696	2.34052	0.99357
4.4	2.69238	4.29756	0.01360	2.35063	0.99787
4.8	3.08534	5.45310	0.00428	2.35421	0.99939
5.2	3.48189	6.76655	0.00115	2.35530	0.99985
5.6	3.88031	8.23899	0.00026	2.35558	0.99997
6.0	4.27964	9.87098	0.00005	2.35564	0.99999
6.4	4.67938	11.66278	0.00001	2.35566	1.00000
6.8	5.07928	13.61452	0.00000	2.35566	1.00000

lue into Eq. (A-4), we obtain numerical values of the dimensionless concentration distribution as shown in the sixth column. Substituting $I_3(\infty)$ into Eq. (4.30), we have the following equation:

$$Sh_x(1-\omega_s) = \frac{Re_x^{1/2}}{2.3556} = 0.424\, Re_x^{1/2}$$

4.3
Numerical Solutions of the Boundary Layer Equations

4.3.1
Quasi-Linearization Method

The boundary layer equations described in the previous sections contain nonlinear terms, and because of this obtaining an analytical solution is usually difficult except in very special cases. However, if an approximate solution can be derived by some other means, we can obtain a numerical solution by a step-by-step iteration technique. Here, we describe a numerical solution of the boundary layer equations known as the *quasi-linearization method*.

If the $(m-1)$-th approximation of the stream function, F_{m-1}, is known, we can approximate the boundary layer equation, Eq. (4.13), by a linear ordinary differential equation of the third order as:

4.3 Numerical Solutions of the Boundary Layer Equations

$$F_m''' + \left(\frac{F_{m-1}}{2}\right) F_m'' = 0 \tag{4.31}$$

where F_m is the m-th approximation of the stream function.

The second- and third-order derivatives of the m-th approximation of the stream functions in finite difference form can be expressed by the following equations:

$$F_{m,j}'' = \frac{1}{2\Delta}(F_{m,j+1}' - F_{m,j-1}') \tag{4.32a}$$

$$F_{m,j}''' = \frac{1}{\Delta^2}(F_{m,j+1}' - 2F_{m,j}' + F_{m,j-1}') \tag{4.32b}$$

Substituting the above equations into Eq. (4.31), we have the following equation:

$$A_j F_{m,j+1}' + B_j F_{m,j}' + C_j F_{m,j-1}' = 0 \tag{4.33}$$

where the coefficients, A_j, B_j, C_j, are given as functions of $F_{m-1,j}$ by the following equation.

$$A_j = \left(\frac{F_{m-1,j}}{4\Delta} + \frac{1}{\Delta^2}\right) \tag{4.34a}$$

$$B_j = -2/\Delta^2 \tag{4.34b}$$

$$C_j = 2/\Delta^2 - A_j \tag{4.34c}$$

Furthermore, if we assume a linear relationship of the stream functions between two successive mesh points, j and $j+1$, we have the following equation:

$$F_{m,j}' = D_j F_{m,j+1}' + E_j \tag{4.35}$$

where the coefficients D_j and E_j are defined as:

$$D_j = \frac{-A_j}{(-2/\Delta^2) + C_j D_{j-1}} \tag{4.36a}$$

$$E_j = \frac{-C_j E_{j-1}}{(-2/\Delta^2) + C_j D_{j-1}} \tag{4.36b}$$

From the boundary conditions shown in Section 4.1, we get the following equations.

$$F_{m,1} = F(0) \tag{4.37a}$$

$$D_1 t = 0 \tag{4.37b}$$

$$E_1 = F'(0) \tag{4.37c}$$

$$F'_{m,N+1} = 1 \tag{4.37d}$$

The m-th approximation of the stream function, F_m, can easily be calculated from the $(m-1)$-th approximation, F_{m-1}, by applying the above equations. Calculation is repeated until convergence of the velocity distribution (= $F'_{m,j}$) is obtained at each mesh point. Depending on the boundary conditions, usually 7 to 12 iterations are necessary, and the calculation is easily performed with Microsoft Excel. Convergence of the stream functions is eventually obtained. A similar algorithm may be applied to calculate the temperature or concentration profiles in a laminar boundary layer along a flat plate.

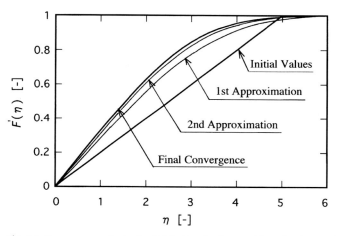

Fig. 4.4 Convergence of velocity distribution by the quasi-linearization method.

Figure 4.4 shows how the convergence of F'_m (m-th approximation of the velocity profile) proceeds with each iteration, where initial values for the velocity profile are tentatively taken as:

$0 \leq \eta \leq 5 \qquad F'(\eta) = 0.2\eta$

$\eta > 5 \qquad F'(\eta) = 1$

4.3.2
Correlation of Heat and Mass Transfer Rates

Figure 4.5 shows the dimensionless concentration distribution in a laminar boundary layer along a flat plate with Schmidt number as the parameter. As stated in Section 4.2, the concentration distribution at $Sc = 1$ coincides with the velocity

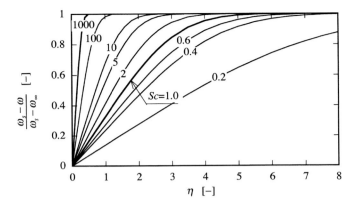

Fig. 4.5 Effect of Schmidt numbers on the concentration distributions in a laminar boundary layer along a flat plate.

distribution in the boundary layer. As the Schmidt number increases from unity, the concentration boundary layer lies within the velocity boundary layer and its thickness becomes smaller and smaller. On the contrary, as the Schmidt number decreases from unity, the concentration boundary layer lies outside of the velocity boundary layer and its thickness becomes larger and larger. The corresponding temperature distribution can easily be obtained by substituting the Schmidt numbers with Prandtl numbers in Fig. 4.5.

Figure 4.6 shows the effect of the Prandtl or Schmidt numbers on the rates of heat or mass transfer in a laminar boundary layer along a flat plate. Numerical data are well correlated by the following equations for the range $0.4 \leq Pr$, $Sc \leq 1000$:

$$Nu_x = 0.332 \, Pr^{1/3} \, Re_x^{1/2} \tag{4.38}$$

$$Sh_x(1 - \omega_s) = 0.332 \, Sc^{1/3} \, Re_x^{1/2} \tag{4.39}$$

The average Nusselt and Sherwood numbers over the entire length of the plate are given by the following equations:

$$\overline{Nu} \equiv \frac{\bar{q}}{\{\kappa(T_s - T_\infty)/L\}} = \int_0^L \frac{0.332 \, Re_x^{1/2} \, Pr^{1/3}}{x} \, dx = 0.664 \, Re^{1/2} \, Pr^{1/3} \tag{4.40}$$

$$\overline{Sh}(1 - \omega_s) \equiv \frac{\overline{N_A}(1 - \omega_s)}{\{\rho D(\omega_s - \omega_\infty)/L\}} = \int_0^L \frac{0.332 \, Re_x^{1/2} \, Sc^{1/3}}{x} \, dx = 0.664 \, Re^{1/2} \, Sc^{1/3} \tag{4.41}$$

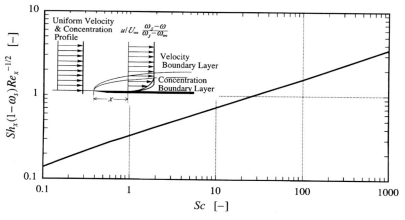

Fig. 4.6 Effect of Prandtl or Schmidt numbers on the rates of heat or mass transfer in a laminar boundary layer along a flat plate.

Example 4.2

Air of relative humidity 80% at 303.15 K and 1 atm is flowing along a flat plate of length 0.2 m and width 0.05 m at a velocity of 0.5 m s^{-1}. Calculate the rate of condensation of water vapor on the plate [kg h^{-1}], if the surface temperature of the condensing wall is 293.15 K.

Solution

The physical properties of the system are as follows:

$p_{303.15K} = 4.24$ kPa

$p = 4.24 \times 0.8 = 3.39$ kPa

$$\omega_\infty = \frac{(3.39)(18)}{\{(101.325 - 3.39)(29) + (3.39)(18)\}} = 2.21 \times 10^{-2}$$

$\rho_\infty = 1.16$ kg m^{-3}, $\mu = 1.86 \times 10^{-5}$ Pa s

We tentatively assume that the surface temperature of the condensate is the same as that of the condensing wall.

$p_{293.15K} = 2.33$ kPa

$$\omega_s = \frac{(2.33)(18)}{\{(101.325 - 2.33)(29) + (2.33)(18)\}} = 0.0144$$

$\rho_s = 1.19$ kg m^{-3}, $\mu_s = 1.83 \times 10^{-5}$ Pa s, $D_{Gs} = 2.48 \times 10^{-5}$ m^2 s^{-1}

The dimensionless numbers for the given conditions are:

$$Sc_s = \frac{(1.83 \times 10^{-5})}{(1.19)(2.48 \times 10^{-5})} = 0.620$$

$$Re = \frac{(1.16)(0.2)(0.5)}{(1.86 \times 10^{-5})} = 6.24 \times 10^3$$

Substituting the above values into Eq. (4.41), we have the following equation:

$$Sh(1 - \omega_s) = (0.664)(0.620)^{1/3}(6.24 \times 10^3)^{1/2} = 44.7$$

from which the mass flux is estimated as:

$$N_A = \frac{(44.7)(1.19)(2.48 \times 10^{-5})(0.0221 - 0.0144)}{(0.2)(1 - 0.0144)}$$

$$= 5.15 \times 10^{-5} \text{ kg m}^{-2} \text{ s}^{-1}$$

The rate of condensation of water vapor is then:

$$= (5.15 \times 10^{-5})(0.2 \times 0.05)(3600)$$

$$= 1.85 \times 10^{-3} \text{ kg h}^{-1}$$

4.4
Mass and Heat Transfer in Extreme Cases

The orders of magnitude of Schmidt numbers of common liquids range from several hundreds to twenty or thirty thousands, while those of liquid metals are 0.04 to 0.002. For extremely large Schmidt numbers or small Prandtl numbers, the thickness of the concentration boundary layer becomes very small in comparison with that of the velocity boundary layer, or the thickness of the thermal boundary layer becomes very large in comparison with that of the velocity boundary layer, and application of a numerical approach in these extreme cases is practically impossible. Some approximate solutions in these extreme cases are discussed below.

4.4.1
Approximate Solutions for Mass Transfer in the Case of Extremely Large Schmidt Numbers

Figure 4.7 shows a schematic representation of the velocity and concentration distributions in a boundary layer along a flat plate in the case of extremely large Schmidt numbers. For the reason described in Section 3.5, the thickness of the

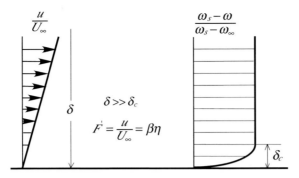

Velocity Distribution   Concentration Distribution

Fig. 4.7 Velocity and concentration distributions in a boundary layer along a flat plate in the case of extremely large Schmidt numbers.

concentration boundary layer, δ_c, is very small in comparison with that of the velocity boundary layer, δ. The velocity distribution in the concentration boundary layer can thus be approximated by the following equation:

$$(u/U_\infty) = F' = \beta\eta \tag{4.42}$$

Integrating Eq. (4.42), we have the following equation for the stream function:

$$F = (\beta/2)\eta^2 \tag{4.43}$$

Substituting the above equation into Eq. (4.29), we obtain the following equation:

$$\left(\frac{d\theta_c}{d\eta}\right)_{\eta=0} = \left\{\int_0^\infty \exp\left(\int_0^\eta -\frac{Sc}{2} F d\eta\right) d\eta\right\}^{-1}$$

$$= \left\{\int_0^\infty \exp\left(-\frac{\beta Sc}{12}\eta^3\right) d\eta\right\}^{-1} = 0.489\,(\beta Sc)^{1/3} \tag{4.44}$$

If we substitute the numerical value of β due to Howarth into Eq. (4.44), we obtain the following approximate solution:

$$Sh_x(1-\omega_s) = 0.339\,Sc^{1/3}\,Re_x^{1/2} \tag{4.45}$$

A similar solution is obtained in the case of extremely large Prandtl numbers:

$$Nu_x = 0.339\,Pr^{1/3}\,Re_x^{1/2} \tag{4.46}$$

4.4.2
Approximate Solutions for Heat Transfer in the Case of Extremely Small Prandtl Numbers [6]

Figure 4.8 shows a schematic representation of the velocity and temperature distributions in a boundary layer along a flat plate in the case of extremely small Prandtl numbers. For the reason described in Section 3.5, the thickness of the thermal boundary layer, δ_T, becomes very large in comparison with that of the velocity boundary layer, δ. The velocity distribution in the thermal boundary layer can thus be approximated by the following equation:

$$\frac{u}{U_\infty} = F' = 1 \tag{4.47}$$

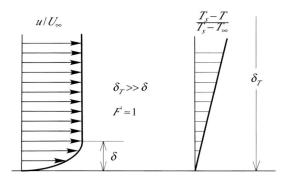

Fig. 4.8 Velocity and temperature distributions in a boundary layer along a flat plate in the case of extremely small Prandtl numbers.

Integrating Eq. (4.47), we have the following equation for the stream function:

$$F = \eta \tag{4.48}$$

Substituting the above equation into Eq. (4.26), we obtain the following equation:

$$\left(\frac{d\theta_T}{d\eta}\right)_{\eta=0} = \left\{\int_0^\infty \exp\left(\frac{-Pr}{4}\eta^2\right) d\eta\right\}^{-1} = \sqrt{\frac{Pr}{\pi}} \tag{4.49}$$

Substituting Eq. (4.49) into Eq. (4.27), we have:

$$Nu_x = 0.560 Pr^{1/2} Re_x^{1/2} \tag{4.50}$$

Equation (4.50) is applicable for heat transfer in the case of extremely small Prandtl numbers, such as for liquid metals and molten salts.

A similar approximate equation can be obtained for mass transfer in the case of extremely small Schmidt numbers:

$$Sh_x(1-\omega_s) = 0.560 Sc^{1/2} Re_x^{1/2} \tag{4.51}$$

However, no materials with such small Schmidt numbers are yet known.

4.5 Effect of an Inactive Entrance Region on Rates of Mass Transfer

4.5.1 Polynomial Approximation of Velocity Profiles and Thickness of the Velocity Boundary Layer

In practical applications, instances in which both the velocity and the concentration boundary layer develop simultaneously from the leading edge are very rare. In most cases, the concentration boundary layer develops from a distance x_0 [m] behind the velocity boundary layer, as shown in Fig. 4.9, and so a numerical solution by quasi-linearization as described in Section 4.3 cannot be applied.

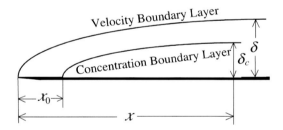

Fig. 4.9 Physical depiction of the effect of an inactive entrance region on the rate of mass transfer in a laminar boundary layer along a flat plate.

Eckert [2] devised an analytical but approximate approach to the effect of an inactive entrance region on heat and mass transfer by assuming the following third-order polynomial approximation for the velocity distribution in the boundary layer:

$$\frac{u}{U_\infty} = a_0 + a_1\left(\frac{y}{\delta}\right) + a_2\left(\frac{y}{\delta}\right)^2 + a_3\left(\frac{y}{\delta}\right)^3 \tag{4.52}$$

where δ is the thickness of the velocity boundary layer.

The boundary conditions are:

$$y = 0: \quad u = v = 0$$
$$y = \delta: \quad u = U_\infty, \quad \frac{\partial u}{\partial y} = 0, \quad \frac{\partial^2 u}{\partial y^2} = 0$$

Equation (4.52) reduces to the following equation:

$$\frac{u}{U_\infty} = \frac{3}{2}\left(\frac{y}{\delta}\right) - \frac{1}{2}\left(\frac{y}{\delta}\right)^3 \tag{4.53}$$

Integrating Eq. (4.8) in terms of y and rearranging the resultant equation by the use of the continuity equation, Eq. (4.7), we have the following equation:

$$\rho \frac{d}{dx} \int_0^\delta (U_\infty - u) u \, dy = \mu \left(\frac{\partial u}{\partial y}\right)_{y=0} \tag{4.54}$$

Substituting Eq. (4.53) into Eq. (4.54), we obtain an ordinary differential equation for the thickness of the velocity boundary layer:

$$\delta \frac{d\delta}{dx} = \frac{280}{13}\left(\frac{\mu x}{\rho U_\infty}\right)^{1/2} \tag{4.55}$$

Integrating Eq. (4.55) in terms of x, we have the following equation for the thickness of the velocity boundary layer:

$$\delta = \left(\frac{280 \mu x}{13 \rho U_\infty}\right)^{1/2} = 4.64 \left(\frac{\mu x}{\rho U_\infty}\right)^{1/2} \tag{4.56}$$

4.5.2
Polynomial Approximation of Concentration Profiles and Thickness of the Concentration Boundary Layer

In a similar way, we can approximate the concentration distribution in the boundary layer by the following third-order polynomial:

$$\theta_c \equiv \frac{\omega_s - \omega}{\omega_s - \omega_\infty} = \frac{3}{2}\left(\frac{y}{\delta_c}\right) - \frac{1}{2}\left(\frac{y}{\delta_c}\right)^3 \tag{4.57}$$

where δ_c is the thickness of the concentration boundary layer.

Integrating Eq. (4.6) in terms of y and rearranging the resultant equation by the use of the continuity equation, Eq. (4.7), we have the following equation:

$$\frac{d}{dx} \int_0^{\delta_c} u(\omega_\infty - \omega) \, dy = D\left(\frac{\partial \omega}{\partial y}\right)_{y=0} \tag{4.58}$$

Substituting Eq. (4.57) into Eq. (4.58), we obtain the following equation for the ratio of the thicknesses of the two boundary layers, that is, the velocity boundary layer and the concentration boundary layer, $\zeta \equiv \delta_c/\delta$:

$$(\omega_s - \omega_\infty) U_\infty \frac{d}{dx}\left\{\delta\left(\frac{3}{20}\zeta^2 - \frac{3}{280}\zeta^4\right)\right\} = \frac{3}{2} D \frac{(\omega_s - \omega_\infty)}{\zeta\delta} \tag{4.59}$$

Assuming that $\zeta < 1$ and $Sc > 1$, and rearranging Eq. (4.59) by the use of Eq. (4.56), we obtain the following equation:

$$\zeta^3 + \frac{4}{3}x\frac{d}{dx}\zeta^3 = \frac{13}{14}\frac{1}{Sc} \tag{4.60}$$

The general solution of the above equation is given by the following equation:

$$\zeta^3 = \frac{13}{14}\frac{1}{Sc} + Cx^{-3/4} \tag{4.61}$$

By use of the boundary condition:

$$x = x_0: \quad \zeta = 0$$

we obtain the following equation for ζ:

$$\zeta \equiv \frac{\delta_c}{\delta} = \left(\frac{13}{14}\right)^{1/3} Sc^{-1/3}\left\{1 - \left(\frac{x_0}{x}\right)^{3/4}\right\}^{1/3} \tag{4.62}$$

Substituting Eq. (4.62) into Eq. (4.30), we obtain the following approximate solution for the local rate of mass transfer in the case of an inactive entrance region of length x_0 [m]:

$$Sh_x(1-\omega_s) \equiv \frac{N_{Ax} x(1-\omega_s)}{\rho D(\omega_s - \omega_\infty)} = \frac{3}{2}\frac{x}{\delta_c} = C_x\{Sh_x(1-\omega_s)\}_0 \tag{4.63}$$

$$C_x = 1.025\left\{1 - \left(\frac{x_0}{x}\right)^{3/4}\right\}^{-1/3} \tag{4.64}$$

where the coefficient, C_x, represents the effect of the inactive entrance region on the local mass transfer rate, and $\{Sh(1-\omega_s)\}_0$ is the dimensionless local diffusional flux for a laminar boundary layer along a flat plate in which both the velocity and concentration boundary layers develop simultaneously:

$$\{Sh(1-\omega_s)\}_0 = 0.332 Sc^{1/3} Re_x^{1/2} \tag{4.39}$$

The effect of the inactive entrance region on the average Sherwood number can easily be obtained by integrating the above equation in terms of x over the entire length of the plate:

$$\overline{Sh}(1-\omega_s) \equiv \frac{\overline{N_A}L}{\rho D(\omega_s - \omega_\infty)} = C_{av}\{\overline{Sh}(1-\omega_s)\}_0 \qquad (4.65)$$

where the coefficient, C_{av}, represents the effect of the inactive entrance length, and $\{\overline{Sh}(1-\omega_s)\}_0$ is the dimensionless average diffusional flux for a laminar boundary layer along a flat plate in which the velocity and concentration boundary layers develop simultaneously:

$$\{\overline{Sh}(1-\omega_s)\}_0 = 0.664 Sc^{1/3} Re^{1/2} \qquad (4.41)$$

Equations similar to Eqs. (4.63) and (4.65) can be obtained for the effect of an inactive entrance region on heat transfer.

Figure 4.10 shows coefficients C_x and C_{av}, which represent correction factors for the effects of an inactive entrance region on the local and average rates of mass transfer.

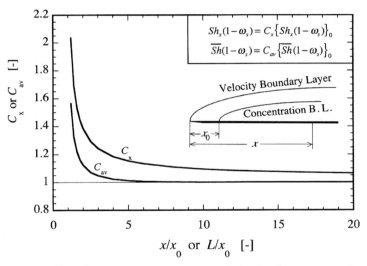

Fig. 4.10 Effect of an inactive entrance region on the local and average rates of mass transfer in a laminar boundary layer along a flat plate.

4.6 Absorption of Gases by a Falling Liquid Film

4.6.1 Velocity Distribution in a Falling Thin Liquid Film According to Nusselt [9]

Figure 4.11 shows a schematic representation of the absorption of gases by a falling liquid film flowing along an inclined flat plate. If we assume that the effect of

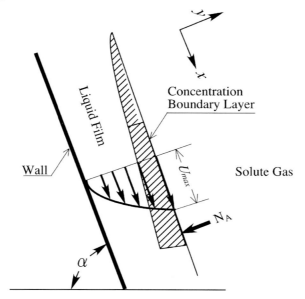

Fig. 4.11 Physical depiction of the absorption of gases by a falling liquid film.

the inertia force is negligibly small, the equation of motion for the falling liquid film can be written as:

$$\rho_L g \sin \alpha + \mu_L \frac{d^2 u}{dy^2} = 0 \tag{4.66}$$

where g is the acceleration due to gravity [m s^{-2}], u is the velocity in the liquid film [m s^{-1}], α is the angle of inclination [rad], δ is the thickness of the liquid film [m], μ_L is the viscosity of the liquid [Pa s], and ρ_L is the density of the liquid [kg m^{-3}].

The boundary conditions are:

$y = 0$: $u = 0$

$y = \delta$: $\partial u / \partial y = 0$

Assuming that the tangential stress at the surface of the liquid film is negligibly small, by integrating Eq. (4.66) with respect to y, we obtain the following equation:

$$u = U_{max} \left\{ 1 - \left(1 - \frac{y}{\delta}\right)^2 \right\} \tag{4.67}$$

$$U_{max} = \frac{\rho_L g \delta^2 \sin \alpha}{2\mu_L} = \frac{3}{2} U_{av} \tag{4.68}$$

where U_{max} is the surface velocity of the liquid film [m s^{-1}]. Equation (4.67) is known as *Nusselt's equation* for the velocity distribution in a falling liquid film.

The following equations are obtained from Eq. (4.67):

The flow rate of the liquid per unit width of the plate, Γ [kg m^{-1} s^{-1}]:

$$\Gamma = \int_0^\delta \rho_L u \, dy = \left(\frac{2\rho_L \delta U_{max}}{3}\right) = \rho_L \delta U_{av} \tag{4.69}$$

The thickness of the liquid film, δ [m]:

$$\delta = \left(\frac{3\mu_L \Gamma}{g\rho_L^2 \sin\alpha}\right)^{1/3} \tag{4.70}$$

The velocity gradient at the wall, β [s^{-1}]:

$$\beta = \left(\frac{du}{dy}\right)_{y=0} = \frac{\rho_L g \delta \sin\alpha}{\mu_L} \tag{4.71}$$

4.6.2
Gas Absorption for Short Contact Times

Figure 4.11 indicates that if the contact time is very short, the concentration distribution in the liquid film will be limited in the region near the interface. Under these circumstances, the diffusion equation can be approximated by the following equation:

$$U_{max}\frac{d\omega}{dx} = D_L \frac{d^2\omega}{dy^2} \tag{4.72}$$

The boundary conditions are:

$x = 0 : \omega = \omega_0$

$y = 0 : \omega = \omega_0$

$y = \delta : \omega = \omega_s$

The contact time of the liquid film at a distance x [m] from the leading edge of the liquid film, t [s], is given by the following equation:

$t = x/U_{max}$

The mass transfer coefficient in this case is easily estimated by the use of *Higbie's penetration model*, as described in Section 3.4.2.

$$k(x) = \sqrt{\frac{D_L}{\pi t}} = \sqrt{\frac{D_L U_{max}}{\pi x}} \tag{4.73}$$

The average mass transfer coefficient over the surface of the liquid film is given by:

$$k_{av} = \frac{1}{L} \int_0^L k(x)\, dx = 1.128 \sqrt{D_L U_{max}/L} \qquad (4.74)$$

Rearranging Eq. (4.74) into dimensionless form, we obtain the following equation:

$$Sh = \frac{k_{av} L}{D_L} = 1.693 \left(\frac{\Gamma}{\rho_L D_L}\right) \sqrt{p} \qquad (4.75)$$

$$p \equiv \left(\frac{D_L L}{\delta^2 U_{max}}\right) \qquad (4.76)$$

Equation (4.74) or (4.75) holds for the absorption of sparingly soluble gases by wetted-wall columns.

4.6.3
Gas Absorption for Long Exposure Times

In the case of intermediate and long exposure times, the thickness of the concentration boundary layer increases and the region occupied by the absorbed gas extends into the region in which a parabolic velocity profile is established. Under these circumstances, the diffusion equation for the liquid film can be expressed as follows:

$$U_{max}\left\{1 - \left(1 - \frac{y}{\delta}\right)^2\right\} \frac{d\omega}{dx} = D_L \frac{d^2\omega}{dy^2} \qquad (4.77)$$

The solution of Eq. (4.77) can generally be expressed by an infinite series expansion, as shown below:

$$\Delta \equiv \frac{(\omega_s - \omega_{av})}{(\omega_s - \omega_0)} = \sum_0^\infty \alpha_j \exp(-\beta_j p) \qquad (4.78)$$

where α_j and β_j are eigenvalues. R. L. Pigford [11] obtained eigenvalues of Eq. (4.78) up to the fourth term, and later W. E. Olbrich and J. W. Wild [10] obtained them up to the tenth term. Table 4.3 shows the eigenvalues of Eq. (4.78) due to Olbrich and Wild.

From the mass balance over the liquid film and substitution of Eq. (4.78) into the resultant equation, we have the following equation:

$$Sh = \left(\frac{\Gamma}{\rho_L D_L}\right) \ln\left(\frac{1}{\Delta}\right) = -\left(\frac{\Gamma}{\rho_L D_L}\right) \ln\left(\sum_0^\infty \alpha_j \exp(-\beta_j p)\right) \qquad (4.79)$$

Table 4.3 Eigenvalues of Eq. (4.78) obtained by W. E. Olbrich & J. W. Wild [10].

j	α	β	j	α	β
1	0.78970	5.1216	6	0.0077	498.09
2	0.09725	36.661	7	0.0055	692.73
3	0.03609	106.249	8	0.0042	919.36
4	0.01868	204.856	9	0.0032	1177.99
5	0.01140	335.473	10	0.0026	1468.63

At long exposure times ($p > 1$), the second term and the higher order terms of p on the right-hand side of Eq. (4.79) become negligibly small in comparison with the first term and we can approximate Eq. (4.79) by the following equation [14]:

$$Sh = 5.12\, p \left(\frac{\Gamma}{\rho_L D_L} \right) \tag{4.80}$$

Figure 4.12 shows the results of calculations of gas absorption using the short contact time model based on Eq. (4.75), the intermediate contact time model based on Eq. (4.79), and the long exposure time model based on Eq. (4.80).

Example 4.3
A mixture of 20% CO_2 and 80% air at a pressure of 1 atm at 293.15 K is absorbed by water at 293.15 K in a wetted-wall column of diameter 20 mm and length

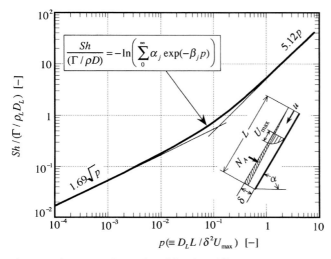

Fig. 4.12 Absorption of gases by a falling liquid film.

200 mm. Calculate the rate of absorption of carbon dioxide [mol h^{-1}] if the flow rate of water is 15.0 kg h^{-1}.

Solution
The physical properties of the system under the specified conditions are:

Henry's constant $= 0.142 \times 10^4$ atm

$\mu_L = 1.00 \times 10^{-3}$ Pa s, $\rho_L = 1000$ kg m^{-3}, $D_L = 1.70 \times 10^{-9}$ m^2 s^{-1},

$c_L = (1000)/(18.0) = 55.6$ kmol m^{-3}

Calculation of liquid film parameters:

$\Gamma = (15.0/3600)/(3.14 \times 0.020) = 6.63 \times 10^{-2}$ kg m^{-1} s^{-1}

$\delta = \{(3)(1.00 \times 10^{-3})(6.63 \times 10^{-2})/(9.81)(1000)^2\}^{1/3} = 2.73 \times 10^{-4}$ m

$U_{max} = (1000)(9.81)(2.73 \times 10^{-4})^2/(2)(1.00 \times 10^{-3}) = 0.365$ m s^{-1}

$p = (1.70 \times 10^{-9})(0.2)/\{(2.73 \times 10^{-4})^2(0.365)\} = 1.25 \times 10^{-2} \ll 1$

Since p is sufficiently small, Eq. (4.74) is applicable to the problem at hand.
Calculation of the absorption rate:

$k_{av} = (1.128)\sqrt{(1.70 \times 10^{-9})(0.365)/(0.2)} = 6.29 \times 10^{-5}$ m s^{-1}

$x_s = (0.2)/(0.142 \times 10^4) = 1.41 \times 10^{-4}$

$N_A = (55.6)(6.29 \times 10^{-5})(1.41 \times 10^{-4}) = 4.93 \times 10^{-7}$ kmol m^{-2} s^{-1}

Surface area for mass transfer $= (3.14)\{0.02 - 2 \times (2.73 \times 10^{-4})\}(0.200)$
$= 1.22 \times 10^{-2}$ m^2

Rate of absorption $= (4.93 \times 10^{-7})(1.22 \times 10^{-2})$
$= 6.00 \times 10^{-9}$ kmol s$^{-1} = 2.16 \times 10^{-2}$ mol h^{-1}

4.7
Dissolution of a Solid Wall by a Falling Liquid Film

H. Kramer and P. J. Kreyger [7] discussed the rate of dissolution of a solid wall by a falling liquid film. In this special case, the Schmidt numbers are very large and we can assume the following relationship:

$\delta \gg \delta_c$

Therefore, the velocity distribution near the wall at which the concentration boundary layer is set up can be approximated by the following equation:

$$u = \beta y \qquad (4.81)$$

4.7 Dissolution of a Solid Wall by a Falling Liquid Film

The diffusion equation can be written as:

$$\beta y \frac{d\omega}{dx} = D_L \frac{d^2\omega}{dy^2} \qquad (4.82)$$

The boundary conditions are:

$x = 0, y > 0: \omega = \omega_0$

$x > 0, y = 0: \omega = \omega_s$

$y = \infty: \omega = \omega_0$

If we apply the following transformations of the variables:

$$X \equiv y \left(\frac{\beta}{9 x D_L}\right)^{1/3} \qquad (4.83a)$$

$$\theta = \frac{\omega - \omega_s}{\omega_0 - \omega_s} \qquad (4.83b)$$

and rearrange Eq. (4.83), we have:

$$\frac{d^2\theta}{dX^2} + 3 X^2 \frac{d\theta}{dX} = 0 \qquad (4.84)$$

The boundary conditions are:

$X = 0: \theta = 0$

$X = \infty: \theta = 1$

The solution of Eq. (4.84) can easily be obtained from:

$$\theta = \frac{\int_0^X \exp(X^3) \, dX}{\int_0^\infty \exp(X^3) \, dX} = \frac{\int_0^X \exp(X^3) \, dX}{0.893} \qquad (4.85)$$

From Eq. (4.85), the rate of dissolution of the wall is given by the following equation:

$$N_{Ax} = -\left(\frac{\rho_L D_L}{1 - \omega_s}\right)\left(\frac{\partial \omega}{\partial y}\right)_{y=0} = 0.538 \, \rho_L D_L \left(\frac{\beta}{x D_L}\right)^{1/3} \left(\frac{\omega_s - \omega_0}{1 - \omega_s}\right) \qquad (4.86)$$

Equation (4.86) can be written in dimensionless form as follows:

$$Sh(1 - \omega_s) = 1.527 \left(\frac{\Gamma}{\rho_L D_L}\right) p^{2/3} \qquad (4.87)$$

Figure 4.13 shows a comparison of Eq. (4.87) with experimental data on the dissolution of benzoic acid in water obtained by Kramer and Kreyger [7].

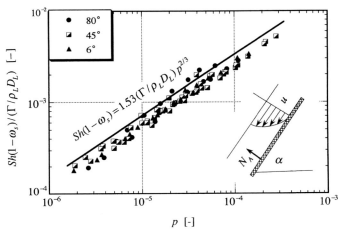

Fig. 4.13 Dissolution of a solid wall by a falling liquid film; comparison of Eq. (4.87) with experimental data obtained by H. Kramer and P. J. Kreyger [7].

4.8
High Mass Flux Effect in Heat and Mass Transfer in Laminar Boundary Layers

4.8.1
High Mass Flux Effect

The velocity distribution in a boundary layer along a flat plate will be much affected by mass injection or suction through the wall, as well as by the friction factor. In a similar way, the temperature and concentration distributions are also affected by mass injection or suction. This phenomenon is known as the *high mass flux effect*. H. von Schu [13] and later J. P. Hartnett and E. R. G. Eckert [4] presented numerical approaches to this problem. Figure 4.14 shows the variation in the velocity distribution in a boundary layer with mass injection or suction, where $F(0)$ is the stream function at the wall:

$$F(0) = -2\left(\frac{v_s}{U}\right) Re_x^{1/2} \qquad (4.88)$$

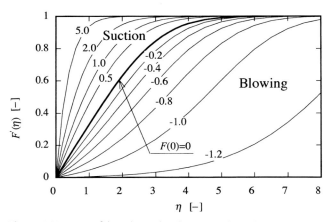

Fig. 4.14 Variation of the velocity distribution in a boundary layer with mass injection or suction.

The figure clearly indicates that the friction factor increases with mass suction and decreases with mass injection. As discussed in Section 2.3, we know that mass flux always accompanies convective mass flux or the normal component of the velocity at the interface, v_s. If the rate of mass transfer reaches a certain high level, the velocity, the temperature, and the concentration distributions in the boundary layer will be affected by the convective mass flux, and the high mass flux effect will be induced. In mass transfer with a phase change, such as the evaporation of volatile liquids or the condensation of vapors in the presence of a non-condensable gas, the rate of mass transfer is usually very large and we cannot neglect this high mass flux effect.

4.8.2
Mickley's Film Model Approach to the High Mass Flux Effect

H. S. Mickley et al. [8] presented a theoretical approach to the high mass flux effect on the basis of the following assumptions:
1) one-dimensional flow;
2) the thickness of the velocity boundary layer, δ, is independent of the rate of mass injection or suction, or the velocity at the wall, v_s [m s^{-1}].

Under these circumstances, the equation of motion can be written as follows:

$$v_s \frac{du}{dy} = \frac{\mu}{\rho} \frac{d^2 u}{dy^2} \qquad (4.89)$$

Substituting the following dimensionless variables into Eq. (4.89):

$$\eta = y/\delta, \quad \beta = u/U_\infty, \quad \Gamma = \rho v_s \delta / \mu$$

the following equation is obtained:

$$\frac{d^2\beta}{d\eta^2} = \Gamma \frac{d\beta}{d\eta} \qquad (4.90)$$

The boundary conditions are:

$$\eta = 0: \quad \beta = 0$$
$$\eta = 1: \quad \beta = 1$$

Integrating Eq. (4.90) using the boundary conditions, we have:

$$\beta = \frac{e^{\Gamma\eta} - 1}{e^{\Gamma} - 1} \qquad (4.91)$$

Differentiating Eq. (4.91) with respect to η, we have:

$$\left(\frac{d\beta}{d\eta}\right)_{\eta=0} = \frac{\Gamma}{e^{\Gamma} - 1} \qquad (4.92)$$

The friction factor is obtained from Eq. (4.92) as:

$$\frac{f_x}{2} = \frac{\mu\left(\frac{du}{dy}\right)_{y=0}}{\rho U_\infty^2} = \left(\frac{\mu}{\rho \delta U_\infty}\right) \frac{\Gamma}{e^{\Gamma} - 1} \qquad (4.93)$$

The friction factor in the absence of mass injection or suction ($\Gamma = 0$), f_0, is given by the following equation:

$$\frac{f_{x0}}{2} = \lim_{\Gamma \to 0}\left(\frac{f}{2}\right) = \left(\frac{\mu}{\rho \delta_0 U_\infty}\right) \qquad (4.94)$$

From Eqs. (4.93) and (4.94), the effect of mass injection or suction on the local friction factor is given by:

$$\frac{f_x}{f_{x0}} = \frac{b}{e^b - 1} = \frac{\ln(1 + B_u)}{B_u} \qquad (4.95)$$

where b and B are the *blowing parameters* defined by the following equations:

$$b \equiv \frac{2}{f_0} \frac{v_s}{U_\infty} \qquad (4.96)$$

$$B_u \equiv \frac{2}{f_x} \frac{v_s}{U_\infty} = b\left(\frac{f_{x0}}{f_x}\right) \tag{4.97}$$

In a similar way, the corresponding equations for the high mass flux effect on heat and mass transfer can be written as follows:

$$\frac{Nu_x}{Nu_{x0}} = \frac{\ln(1 + B_H)}{B_H} \tag{4.98}$$

$$\frac{Sh_x(1-\omega_s)}{Sh_{x0}(1-\omega_s)} = \frac{\ln(1 + B_M)}{B_M} \tag{4.99}$$

where Nu_{x0} and sh_{x0} are the local Nusselt number and the Sherwood number in the absence of mass injection and suction, respectively, and B_H and B_M are the *transfer numbers* for heat and mass transfer defined by the following equations:

$$B_H \equiv \left(\frac{v_s}{U_\infty}\right)\left(\frac{RePr}{Nu}\right) \tag{4.100}$$

$$B_M \equiv \left(\frac{v_s}{U_\infty}\right)\left(\frac{ReSc}{Sh(1-\omega_s)}\right) \tag{4.101}$$

4.8.3
Correlation of High Mass Flux Effect for Heat and Mass Transfer

Figure 4.15 shows numerical results for the high mass flux effect on mass transfer in a laminar boundary layer along a flat plate obtained by the quasi-linearization

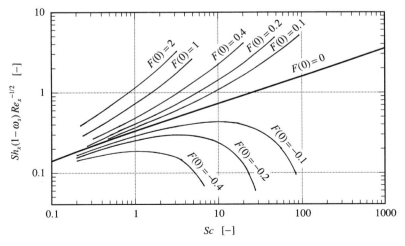

Fig. 4.15 High mass flux effect on mass transfer.

method. The bold solid line in the figure represents numerical solutions without mass injection or suction ($F(0) = 0$), i.e., in the ordinary boundary layer, while the thin solid lines represent numerical solutions at constant $F(0)$. Similar results are obtained for heat transfer.

The figure indicates that the high mass flux effect is a complicated function of Schmidt numbers (or Prandtl numbers for heat transfer) and mass injection or suction.

However, relatively simple correlations can be obtained for the high mass flux effect on heat and mass transfer if we use correlations in terms of transfer numbers.

Heat transfer:

$$Nu_x = 0.332\, Re_x^{1/2}\, Pr^{1/3}\, g\,(B_H) \tag{4.102}$$

$$g\,(B_H) \equiv \frac{Nu_x}{Nu_{x0}} = \frac{1}{0.09 + 0.91\,(1 + B_H)^{0.8}} \tag{4.103}$$

Mass transfer:

$$Sh_x\,(1 - \omega_s) = 0.332\, Re_x^{1/2}\, Sc^{1/3}\, g\,(B_M) \tag{4.104}$$

$$g\,(B_M) \equiv \frac{Sh_x\,(1 - \omega_s)}{Sh_{x0}\,(1 - \omega_s)} = \frac{1}{0.09 + 0.91\,(1 + B_M)^{0.8}} \tag{4.105}$$

Similar equations hold for the average Nusselt numbers, $\overline{Nu}$, and the dimensionless diffusional fluxes, $\overline{Sh}\,(1 - \omega_s)$:

$$\overline{Nu} = 0.664\, Re^{1/2}\, Pr^{1/3}\, g\,(B_H) \tag{4.106}$$

$$\overline{Sh}\,(1 - \omega_s) = 0.664\, Re^{1/2}\, Sc^{1/3}\, g\,(B_M) \tag{4.107}$$

Figure 4.16 shows the final correlation of the high mass flux effect in terms of transfer number B ($B = B_H$ for heat transfer, $B = B_M$ for mass transfer). The bold solid line in the figure represents the correlation of numerical solutions for the effect of high mass flux on heat and mass transfer (Eqs. (4.103) and (4.105)), while the dash-dotted line represents the analytical solution obtained by Mickley's film model (Eqs. (4.98) and (4.99)). The difference between the two solutions may be due to the fact that the film model is based on the unrealistic assumption that the thickness of the boundary layer does not change with mass injection or suction. This means that the film model is only applicable within a small range of transfer numbers.

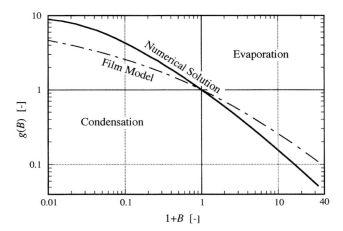

Fig. 4.16 Correlation of high mass flux effects on heat and mass transfer.

Rearranging the transfer numbers for mass transfer, we obtain the following equation:

$$B_M \equiv \left(\frac{v_s}{U_\infty}\right)\left(\frac{ReSc}{Sh(1-\omega_s)}\right) = \left(\frac{v_s}{U_\infty}\right)\left(\frac{\rho L U_\infty}{\mu}\right)\left(\frac{\mu}{\rho D}\right)\left(\frac{\rho D(\omega_s - \omega_\infty)}{N_A(1-\omega_s)L}\right)$$

$$= \left(\frac{\rho v_s}{N_A}\right)\left(\frac{\omega_s - \omega_\infty}{1-\omega_s}\right) = \frac{\omega_s - \omega_\infty}{1-\omega_s} \tag{4.108}$$

Equation (4.108) indicates that the transfer number for mass transfer, B_M, is a function of the interface (surface) and the free stream concentrations. This fact is particularly pertinent in dealing with important practical problems such as condensation in the presence of a non-condensable gas or the evaporation of a volatile liquid, detailed applications of which are discussed in Chapter 9.

Transfer numbers for heat transfer can only be transformed into the following equation if wet-bulb temperature conditions, ($q_G = l_v N_A$), hold at the interface:

$$B_H \equiv \left(\frac{v_s}{U_\infty}\right)\left(\frac{RePr}{Nu}\right) = \left(\frac{v_s}{U_\infty}\right)\left(\frac{\rho L U_\infty}{\mu}\right)\left(\frac{c_p \mu}{\kappa}\right)\left(\frac{\kappa (T_s - T_\infty)}{q_G L}\right)$$

$$= \frac{\rho v_s c_p (T_s - T_\infty)}{q_G} = \frac{c_p (T_s - T_\infty)}{l_v} \tag{4.109}$$

Here, c_p is the specific heat at constant pressure [J kg^{-1} K^{-1}], l_v is the latent heat of vaporization [J kg^{-1}], T is the temperature [K], U_∞ is the free stream velocity [m s^{-1}], v_s is the normal component of the velocity at the interface [m s^{-1}], and the subscript 0 denotes heat or mass flux without mass injection or suction.

Example 4.4

A mixture of 30% hexane and 70% air at a temperature of 303.15 K and a total pressure of 1 atm is supplied to a vertical flat plate condenser of length 0.2 m and width 0.05 m at a velocity of 0.5 m s^{-1}. Calculate the rate of condensation of vapor [kg h^{-1}] if the temperature of the wall is 294 K.

Solution

Assuming that the surface temperature of the condensate is equal to that of the wall, the physical properties of the system are as follows:

$\omega_\infty = 0.560$

$\rho_{G\infty} = 1.86$ kg m^{-3}, $\mu_{G\infty} = 1.15 \times 10^{-5}$ Pa s,

$p_s = 16.83$ kPa, $\omega_s = 0.372$,

$\rho_{Gs} = 1.60$ kg m^{-3}, $\mu_{Gs} = 1.34 \times 10^{-5}$ Pa s, $D_{Gs} = 7.48 \times 10^{-6}$ m^2 s^{-1}

The Reynolds and Schmidt numbers in this case are as follows:

$$Re = \frac{(1.86)(0.5)(0.2)}{(1.15 \times 10^{-5})} = 1.62 \times 10^4$$

$$Sc = \frac{(1.34 \times 10^{-5})}{(1.60)(7.48 \times 10^{-6})} = 1.12$$

As in the first approximation, we calculate the rate of mass transfer using Eq. (4.41), that is, without considering the high mass flux effect:

$$\overline{Sh}_0 (1 - \omega_s) = (0.664)(1.12)^{1/3}(1.62 \times 10^4)^{1/2} = 87.8$$

$$\{N_A\}_0 = \frac{(87.8)(1.60)(7.48 \times 10^{-6})(0.560 - 0.372)}{(0.2)(1 - 0.372)}$$

$$= 1.57 \times 10^{-3} \text{ kg m}^{-2} \text{ s}^{-1}$$

The transfer number under the given conditions:

$$B_M = \frac{(0.372 - 0.560)}{(1 - 0.372)} = -0.299$$

The high mass flux effect can then be calculated by Eq. (4.105):

$$g(B_M) = \frac{1}{\{0.09 + 0.91(1 - 0.299)^{0.8}\}} = 1.29$$

This indicates that the net rate of condensation will be 1.29 times higher than the calculated rate without the high mass flux effect. The net condensation rate will be as follows:

$$\text{Rate of condensation} = (1.29)(1.57 \times 10^{-3})(0.2 \times 0.05)(3600)$$
$$= 7.29 \times 10^{-2} \text{ kg h}^{-1}$$

In this example, we have presented an approximate solution in which the surface temperature is tentatively assumed. The calculation proceeds in a straightforward manner once the surface temperature is assigned. A more general solution, in which the surface temperature is determined by the heat balance between the liquid and the vapor, will be discussed in Chapter 9.

References

1 K. Asano, Y. Nakano, and M. Inaba, "Forced Convection Film Condensation of Vapors in the Presence of Noncondensable Gas on a Small Vertical Flat Plate", *Journal of Chemical Engineering of Japan*, **12**, [3], 196–202 (1979).

2 E. R. G. Eckert and R. M Drake, Jr., "*Heat and Mass Transfer*", pp. 139–142, 173–176, McGraw-Hill (1959).

3 V. M. Falkner and S. W. Skan, "Solutions of the Boundary Layer Equations", *Philosophical Magazine and Journal of Science*, **12**, [80], 865–896 (1931).

4 J. P. Hartnett and E. R. G. Eckert, "Mass Transfer Cooling in a Laminar Boundary Layer with Constant Fluid Properties", *Transactions of the American Society of Mechanical Engineers*, **79**, 247–254 (1957).

5 L. Howarth, "On the Solution of the Laminar Boundary Layer Equations", *Proceedings of the Royal Society* (London), **A164**, 547–579 (1938).

6 W. M. Kays and M. E. Crawford, "*Convective Mass Transfer*", 2nd ed., pp. 133–150, McGraw-Hill (1980).

7 H. Kramer and P. J. Kreyger, "Mass Transfer between a Flat Surface and a Falling Liquid Film", *Chemical Engineering Science*, **6**, 42–48 (1956).

8 H. S. Mickley, R. C. Ross, A. L. Squyers, and W. E. Stewart, *NACA Technical Note*, **3208** (1954).

9 W. von Nusselt, "Die Oberflächenkondensation des Wasserdampfes", *Zeitschrift des Vereines Deutscher Ingenieure*, **60**, [27], 541–546 (1916).

10 W. E. Olbrich and J. W. Wild, "Diffusion from the Free Surface into a Liquid Film in Laminar Flow over Defined Shapes", *Chemical Engineering Science*, **24**, 25–32 (1956).

11 R. L. Pigford, Ph.D. Thesis in Chemical Engineering, University of Illinois (1942).

12 H. Schlichting, "*Boundary Layer Theory*", 6th ed., pp. 117–134, McGraw-Hill (1968).

13 H. von Schu, "Über die Lösung der laminaren Grenzsichtgleichung an der ebenen Platte für Geschwindigkeits und Temperaturfeld bei veränderlichen Stoffwerten und für das Diffusionfeld bei höheren Konzentrationen", *Zeitschrift für Angewandte Mathematik und Mechanik*, **25/27**, [2], 54–60 (1951).

14 T. K. Sherwood, R. L. Pigford, and C. R. Wilke, "*Mass Transfer*", pp. 205–211, McGraw-Hill (1975).

5
Heat and Mass Transfer in a Laminar Flow inside a Circular Pipe

5.1
Velocity Distribution in a Laminar Flow inside a Circular Pipe

Heat and mass transfer in a laminar flow inside a circular pipe are of fundamental importance for the design of shell and tube exchangers, and for absorption, evaporation, and distillation in wetted-wall columns. Figure 5.1 shows a schematic representation of the axial variation of the velocity distribution in a circular pipe. A uniform velocity distribution at the entrance region is soon deformed due to the effect of viscosity near the wall, and a velocity boundary layer develops in the downstream region of the pipe. The thickness of the velocity boundary layer increases with increasing distance from the entrance and finally a fully developed velocity distribution is established far away from the entrance. The velocity distribution shows no further change thereafter.

For a fully developed flow in a circular pipe, the flow is axially symmetric and the equation of motion can be written as follows:

$$-\frac{\partial P}{\partial z} + \frac{\mu}{r}\frac{\partial}{\partial r}\left(r\frac{\partial v_z}{\partial r}\right) = 0 \tag{5.1}$$

The boundary conditions are:

$$r = R = D_T/2: \quad v_z = 0 \tag{5.2a}$$

$$r = 0: \quad \partial v_z/\partial r = 0 \tag{5.2b}$$

Integrating Eq. (5.1) with respect to r using the above boundary conditions, we obtain the following well-known velocity distribution for a laminar flow inside a circular pipe, a parabolic velocity distribution:

$$v_z = 2U_m\left\{1 - \left(\frac{r}{R}\right)^2\right\} \tag{5.3}$$

Mass Transfer. From Fundamentals to Modern Industrial Applications. Koichi Asano
Copyright © 2006 WILEY-VCH Verlag GmbH & Co. KGaA, Weinheim
ISBN: 3-527-31460-1

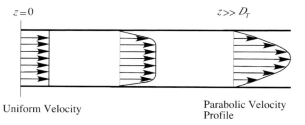

Uniform Velocity   Parabolic Velocity Profile

Fig. 5.1 Axial variation of the velocity distribution in a laminar flow inside a circular pipe.

The average velocity over the cross-section of the pipe is given by:

$$U_m = \frac{1}{\pi R^2} \int_0^R 2\pi r v_z \, dr = \frac{D_T^2}{32\mu} \frac{\Delta P}{L} \tag{5.4}$$

where D_T is the inner diameter of the pipe [m], L is the total length of the pipe [m], ΔP is the pressure drop of the fluid along the axis of the pipe of length L [Pa], U_m is the average velocity over the cross-section of the pipe [m s^{-1}], ρ is the density of the fluid [kg m^{-3}], and μ is the viscosity of the fluid [Pa s].

From Eq. (5.3), the friction factor is given by the following equation:

$$f = \frac{-\mu \left(\frac{\partial v_z}{\partial r}\right)_{r=R}}{(\rho U_m^2/2)} = \frac{16}{Re} \tag{5.5}$$

where Re is the Reynolds number defined by the following equation:

$$Re \equiv \rho D_T U_m / \mu \tag{5.6}$$

Since the transition from laminar to turbulent flow takes place at around $Re = 2300$, Eq. (5.5) is applicable over the range in which the Reynolds number is less than 2300. For this reason, the discussion in this chapter is restricted to the range $Re \leq 2300$.

5.2
Graetz Numbers for Heat and Mass Transfer

5.2.1
Energy Balance over a Small Volume Element of a Pipe

Figure 5.2 shows the energy balance over a small volume element of a pipe. The energy balance over the volume element can be expressed by the following equation:

5.2 Graetz Numbers for Heat and Mass Transfer

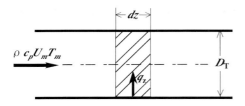

Fig. 5.2 Energy balance over a small volume element of a pipe.

$$\pi D_T \, dz q_z = \frac{\pi D_T^2}{4} \rho c_p U_m \, dT_m \tag{5.7}$$

where c_p is the specific heat of the fluid at constant pressure [J kg^{-1} K^{-1}], D_T is the inner diameter of the pipe [m], q_z is the local heat flux at z [W m^{-2}], U_m is the average velocity of the fluid over the cross-section [m s^{-1}], and ρ is the density of the fluid [kg m^{-3}]. T_m is the average temperature of the fluid over the cross-section of the pipe at z [K], defined by the following equation:

$$T_m \equiv \frac{1}{\pi R^2 U_m} \int_0^R 2\pi r u T \, dr \tag{5.8}$$

Rearranging Eq. (5.7), we have the following equation for the local wall heat flux:

$$q_z = \left(\frac{\rho c_p D_T U_m}{4} \right) \frac{dT_m}{dz} \tag{5.9}$$

Rearranging Eq. (5.9) into dimensionless form, we obtain the following equation for the local Nusselt number:

$$Nu_{loc} \equiv \frac{q_z D_T}{\kappa (T_w - T_m)} = \left(\frac{\rho c_p D_T^2 U_m}{4\kappa} \right) \frac{1}{(T_w - T_m)} \frac{dT_m}{dz} \tag{5.10}$$

where κ is the thermal conductivity of the fluid [W m^{-1} K^{-1}].

Integrating Eq. (5.10) with respect to z, the average Nusselt number over the inner surface of the pipe of length L is given by the following equation:

$$\overline{Nu} \equiv \frac{\bar{q} D_T}{\kappa \Delta T_{lm}} = \frac{Gz_H}{\pi} \ln \left(\frac{T_w - T_0}{T_w - \overline{T}_{out}} \right) \tag{5.11}$$

where Gz_H is the *Graetz number for heat transfer* defined by the following equation:

$$Gz_H \equiv \frac{\pi \rho c_p D_T^2 U_m}{4 \kappa L} = \left(\frac{\pi}{4} \right) \left(\frac{D_T}{L} \right) (Re Pr) \tag{5.12}$$

ΔT_{lm} is the logarithmic mean temperature driving force defined by the following equation:

$$\Delta T_{lm} \equiv \frac{(T_w - T_0) - (T_w - \overline{T}_{out})}{\ln\left(\dfrac{T_w - T_0}{T_w - \overline{T}_{out}}\right)} \tag{5.13}$$

and $\bar{q}$ is the average heat flux over the length of the pipe [W m^{-2}]; T_0, $\overline{T}_{out}$, and T_w are the inlet (initial) temperature, the average outlet temperature, and the wall temperature [K], respectively.

5.2.2
Material Balance over a Small Volume Element of a Pipe

If we apply material balance over a small volume element of length dz, we obtain the following equation for the diffusional flux:

$$Sh_{loc}\,(J_A/N_A) \equiv \frac{J_A D_T}{\rho D(\omega_s - \omega_m)} = \left(\frac{D_T^2 U_m}{4D}\right)\frac{1}{\omega_s - \omega_m}\frac{d\omega_m}{dz} \tag{5.14}$$

where D is the diffusivity [m^2 s^{-1}], J_A is the local diffusional flux of component A at z [kg m^{-2} s^{-1}], N_A is the local mass flux of component A at z [kg m^{-2} s^{-1}], ω_s is the surface concentration of component A (equilibrium concentration), and ω_m is the average concentration over the cross-section at z, defined by the following equation:

$$\omega_m \equiv \frac{1}{\pi R^2 U_m} \int_0^R 2\pi r u \omega \, dr \tag{5.15}$$

The dimensionless diffusional flux over the entire length of the pipe is given by the following equation:

$$\overline{Sh}\,(\overline{J_A}/\overline{N_A}) \equiv \frac{\overline{J_A} D_T}{D\Delta\omega_{lm}} = \frac{Gz_M}{\pi} \ln\left(\frac{\omega_s - \omega_0}{\omega_s - \omega_{out}}\right) \tag{5.16}$$

where Gz_M is the Graetz number for mass transfer defined by the following equation:

$$Gz_M \equiv \frac{\pi D_T^2 U_m}{4DL} = \left(\frac{\pi}{4}\right)\left(\frac{D_T}{L}\right)(Re Sc) \tag{5.17}$$

and $\Delta\omega_{lm}$ is the logarithmic mean concentration driving force defined by the following equation:

$$\Delta \omega_{lm} \equiv \frac{(\omega_s - \omega_0) - (\omega_s - \omega_{out})}{\ln\left(\dfrac{\omega_s - \omega_0}{\omega_s - \omega_{out}}\right)} \tag{5.18}$$

where ω_0 [–], ω_{out}, and ω_s [–] are the inlet (initial) concentration, the average outlet concentration, and the surface concentration (equilibrium concentration), respectively.

5.3
Heat and Mass Transfer near the Entrance Region of a Circular Pipe

5.3.1
Heat Transfer near the Entrance Region at Constant Wall Temperature

The velocity distribution near the entrance region of a circular pipe is generally uniform. As a result, the energy equation for this region can be written as follows:

$$\rho c_p U \frac{\partial T}{\partial z} = \kappa \left(\frac{\partial^2 T}{\partial r^2} + \frac{1}{r} \frac{\partial T}{\partial r} \right) \tag{5.19}$$

Partial differential equations akin to Eq. (5.19) are commonly known as heat conduction equations, the solutions of which are applied in many fields of science and engineering.

The boundary conditions are:

$$r = R(= D_T/2): \quad T = T_w \tag{5.20a}$$

$$z = 0: \quad T = T_0 \tag{5.20b}$$

Although detailed discussions on the general nature of the above equation are presented in a standard text [1], an infinite series expansion solution for the above boundary conditions can be written as follows:

$$\frac{T_w - T}{T_w - T_0} = 2 \sum_{n=1}^{\infty} \frac{J_0(2 r \alpha_n / D_T)}{J_1(\alpha_n)} \frac{1}{\alpha_n} \exp\left(\frac{-\pi}{Gz_H} \alpha_n^2 \right) \tag{5.21}$$

where J_0 and J_1 are zeroth- and first-order Bessel functions of the first kind, and α_n are the eigenvalues of the zeroth-order Bessel function of the first kind.

From Eq. (5.21), the average temperature over the cross-section of the pipe, T_m [K], is given by the following equation:

$$\frac{T_w - T_m}{T_w - T_0} \equiv \frac{4}{\pi D_T^2} \int_0^{D_T/2} 2 \pi r \left(\frac{T_w - T}{T_w - T_0} \right) dr = 4 \sum_{n=1}^{\infty} \frac{1}{\alpha_n^2} \exp\left(\frac{-\pi \alpha_n^2}{Gz_H} \right) \tag{5.22}$$

Substituting Eq. (5.22) into Eq. (5.11), we obtain the following equation for the average Nusselt number over the entire length of the pipe:

$$\overline{Nu} = \frac{Gz_H}{\pi} \ln \left[4 \sum_{n=1}^{\infty} \frac{1}{\alpha_n^2} \exp\left(\frac{-\pi\alpha_n^2}{Gz_H}\right) \right]^{-1} \quad (5.23)$$

Table 5.1 shows the first ten eigenvalues of the zeroth-order Bessel function of the first kind.

Table 5.1 Eigenvalues of the zeroth-order Bessel function of the first kind.

n	α_n	n	α_n
1	2.40483	6	18.07106
2	5.52008	7	21.21164
3	8.65373	8	24.35247
4	11.79153	9	27.49348
5	14.93092	10	30.63461

5.3.2
Mass Transfer near the Entrance Region at Constant Wall Concentration

In a similar way, the dimensionless average diffusional flux over the entire length of a pipe with uniform velocity distribution is given by the following equation:

$$\overline{Sh}\,(\overline{J_A}/N_A) = \frac{Gz_M}{\pi} \ln \left[4 \sum_{n=1}^{\infty} \frac{1}{\alpha_n^2} \exp\left(\frac{-\pi\alpha_n^2}{Gz_M}\right) \right]^{-1} \quad (5.24)$$

Under normal operating conditions, gas/liquid in contact with sieve-tray columns is usually not in a bubbling regime but in a channeling regime. Under such circumstances, mass transfer may be adequately described by a model made up of a group of short circular pipes in which the distribution of gas velocity is uniform. H. Kosuge and K. Asano [5] presented a simulation approach to mass transfer in multicomponent distillation by sieve-tray columns by the use of Eq. (5.24). They reported fairly good agreement between the simulation and the observed data for the ternary distillation of hexane/methylcyclopentane/benzene systems, as reported by Miskin et al., and of cyclohexane/heptane/toluene systems, as reported by Medina et al.

5.4
Heat and Mass Transfer in a Fully Developed Laminar Flow inside a Circular Pipe

5.4.1
Heat Transfer at Constant Wall Temperature

The velocity distribution inside a circular pipe in the region far away from the entrance is fully developed and a parabolic velocity distribution is established. The energy equation for this region can be written as follows:

$$\left\{1-\left(\frac{r}{R}\right)^2\right\}\frac{\partial T}{\partial z} = \left(\frac{\kappa}{2\rho c_p U_m}\right)\left(\frac{\partial^2 T}{\partial r^2} + \frac{1}{r}\frac{\partial T}{\partial r}\right) \tag{5.25}$$

The boundary conditions are:

$$r = R(D_T/2): \quad T = T_w \tag{5.20a}$$

$$z = 0: \qquad T = T_0 \tag{5.20b}$$

The solution of Eq. (5.25) is known as *Graetz's problem*, in honor of L. Graetz [2] for his pioneering work on a theoretical approach to laminar heat transfer inside a circular pipe. Because of the nonlinear nature of the equation, obtaining a perfect infinite series expansion solution, as in the case of a uniform velocity distribution, is rather difficult.

The solution of Eq. (5.25) is known to be of the form shown by the following equation:

$$\frac{T_w - T_m}{T_w - T_0} = 8\sum_{n=1}^{\infty}\frac{G_n}{\lambda_n^2}\exp\left(\frac{-\pi\lambda_n^2}{2Gz_H}\right) \tag{5.26}$$

A major difficulty with Graetz's problem is obtaining sufficient numbers of eigenvalues, λ_n, and coefficients, G_n. For a long time, only the first three or four eigenvalues had been obtained. J. R. Sellars, M. Tribus, and J. S. Klein [6] succeeded in obtaining asymptotic expressions for higher eigenvalues. Table 5.2 shows the results of numerical calculations on eigenvalues, λ_n, and coefficients, G_n, of Eq. (5.26). Readers interested in detailed discussions on the solution of Eq. (5.25) are referred to coverage elsewhere [3, 6].

Substituting Eq. (5.26) into Eq. (5.11), we obtain the following equation for the average Nusselt number over the length of the circular pipe:

$$\overline{Nu} = \frac{Gz_H}{\pi}\ln\left[8\sum_{n=1}^{\infty}\frac{G_n}{\lambda_n^2}\exp\left(\frac{-\pi\lambda_n^2}{2\,Gz_H}\right)\right]^{-1} \tag{5.27}$$

Table 5.2 Eigenvalues λ_n and coefficients G_n of Eq. (5.26) obtained by J. R. Sellars, M. Tribus, and J. S. Klein [6].

n	λ_n	G_n
1	7.312	0.749
2	44.62	0.544
3	113.8	0.463
4	215.2	0.414
5	348.5	0.382
$n > 3$	$\lambda_n = 4(n-1) + 8/3$	$G_n = 1.01276\, \lambda_n^{-1/3}$

Although numerical evaluation of the right-hand side of Eq. (5.25) is laborious and tedious, the calculation is made relatively simply and with sufficient accuracy by the use of the following approximate equations [7]:

Hausen's approximation:

$Gz_H \leq 60$:

$$\overline{Nu} = 3.66 + \frac{0.085\, Gz_H}{1 + 0.047\, Gz_H^{2/3}} \tag{5.28}$$

Leveque's approximation:

$Gz_H \geq 60$:

$$\overline{Nu} = 1.65\, Gz_H^{1/3} \tag{5.29}$$

5.4.2
Mass Transfer at Constant Wall Concentration

The solution of Graetz's problem for mass transfer under constant surface concentration conditions can be obtained in a similar manner as the corresponding solution for heat transfer. The dimensionless average diffusional flux over the length of a pipe is given by the following equation:

$$\overline{Sh}\,(\overline{J_A/N_A}) = \frac{Gz_M}{\pi} \ln\left[8 \sum \frac{G_n}{\lambda_n^2} \exp\left(\frac{-\pi \lambda_n^2}{2 Gz_M}\right)\right] \tag{5.30}$$

Numerical values of the right-hand side of Eq. (5.30) can be obtained by reference to Tab. 5.2. Application of Eq. (5.28) or Eq. (5.29) instead of Eq. (5.30) is also possible by substituting Gz_H with Gz_M.

Figure 5.3 shows a comparison of the two theoretical equations for heat and mass transfer inside a circular pipe in the case of a uniform velocity distribution (entrance region) and that for the case of a parabolic velocity distribution (fully developed flow), where the Nusselt numbers are shown as a function of Graetz

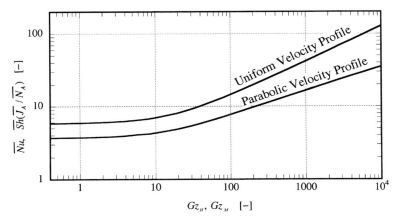

Fig. 5.3 Heat and mass transfer inside a circular pipe.

number for heat transfer ($Gz_H \equiv \pi \rho c_p D_T^2 U_m / 4\kappa L$) or the dimensionless diffusional flux, $\overline{Sh}\,(\overline{J_A}/\overline{N_A})$, as a function of Graetz number for mass transfer ($Gz_M \equiv \pi D_T^2 U_m / 4DL$).

5.5
Mass Transfer in Wetted-Wall Columns

The evaporation of pure liquids and the absorption of easily soluble gases by wetted-wall columns are phenomena for which the gas-phase mass transfer resistances are rate-controlling. Binary and multicomponent distillations are generally gas-phase mass transfer resistance controlled, as will be discussed in Chapter 10. In these cases, Eq. (5.30) is applicable.

H. Kosuge and K. Asano [4] presented an experimental approach to the ternary distillation of the acetone/methanol/ethanol system in wetted-wall columns. Figure 5.4 shows a comparison of their experimental data with the theoretical values obtained using Eqs. (5.24) and (5.30). The ordinate is the dimensionless diffusional flux of component i, $\overline{Sh_{G,i}}\,(\overline{J_i}/\overline{N_i})$, corrected for the effect of partial condensation of vapor, $g_i\,(Re_w, Sc_{G,i})$, while the abscissa is the Graetz number for multicomponent mass transfer, $Gz_{M,i}(= (\pi D_T^2 U_m / 4 D_i L))$. The fact that the data for all of the components for various column lengths showed good agreement with Eq. (5.30) is quite remarkable. On the other hand, on comparing mass flux data, a poor correlation was observed. This indicates that the effect of convective mass flux is considerable in distillation.

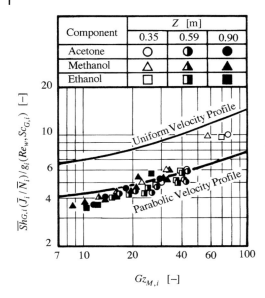

Fig. 5.4 Comparison of Eqs. (5.24) and (5.30) with ternary distillation data obtained by H. Kosuge & K. Asano [4].

Example 5.1

Water at 20 °C is supplied to a wetted-wall column of length 0.6 m and inner diameter 0.02 m at a rate of 0.005 kg m^{-1} s^{-1}. Dry air at 20 °C and atmospheric pressure is supplied counter-currently from the bottom of the column. If the velocity of the air is 1.7 m s^{-1}, calculate the rate of evaporation of water [kg h^{-1}] assuming that the flow of air is fully developed from the bottom of the column and that the surface temperature of the water is 20 °C.

Solution

The physical properties of the system under the given conditions are as follows:

$\mu_{Air} = 1.81 \times 10^{-5}$ Pa s

$\rho_{Air} = 1.20$ kg m^{-3}

$D_{H_2O} = 2.40 \times 10^{-5}$ m^2 s^{-1}

$p_s = 2.33$ kPa

$y_s = (2.33/101.325) = 0.0230$

$\omega_s = \dfrac{(0.0230)(18)}{(0.0230)(18) + (0.977)(29)} = 0.0144$

The Reynolds number is given by:

$Re = \dfrac{(1.2)(0.02)(1.7)}{(1.81 \times 10^{-5})} = 2266 < 2300$

5.5 Mass Transfer in Wetted-Wall Columns

This indicates that the flow in the column is a laminar one. We will assume that the equation for a fully developed flow is applicable to the present problem.

The Graetz number for mass transfer:

$$Gz_M = \frac{(0.785)(1.7)(0.02)}{(2.40 \times 10^{-5})}\left(\frac{0.02}{0.60}\right) = 37.1$$

Substituting the above value into Eq. (5.28) by substituting Gz_H with Gz_M, we have:

$$\frac{N_A D_T}{\rho D \Delta \omega_{lm}}(\overline{J_A/N_A}) = 3.66 + \frac{(0.085)(37.1)}{1 + (0.047)(37.1)^{2/3}} = 5.73 \quad (A)$$

Since the evaporation of a liquid is unidirectional diffusion, the ratio of diffusion to mass flux can be calculated by the use of Eq. (2.21) as:

$$(\overline{J_A/N_A}) = (1 - 0.0144) = 0.986$$

The left-hand side of Eq. (A) includes the two unknown variables, N_A and $\Delta\omega_{lm}$, and so direct solution of Eq. (A) is not possible.

A trial-and-error solution by assuming the outlet concentration $\overline{\omega_2}$ is thus carried out.

Surface area:

$$S = (3.14)(0.02)(0.6) = 3.77 \times 10^{-2} \text{ m}^2$$

Concentration driving force:

$$\Delta\omega_1 = 0.0144, \quad \Delta\omega_2 = 0.0144 - \overline{\omega_2}$$

$$\Delta\omega_{lm} \approx \frac{(\Delta\omega_1 + \Delta\omega_2)}{2} = 0.0144 - 0.5\overline{\omega_2} \quad (B)$$

Flow rate at the exit:

$$G_{Air} = (0.785)(0.02)^2(1.2)(1.7) = 6.41 \times 10^{-4} \text{ kg s}^{-1}$$

$$G_{Water} = N_A S = \frac{(5.73)(1.2)(2.40 \times 10^{-5})}{(0.02)(0.986)}(3.77 \times 10^{-2})\Delta\omega_{lm}$$

$$= 3.15 \times 10^{-4} \Delta\omega_{lm} \quad (C)$$

Outlet vapor concentration:

$$\overline{\omega_2} = \frac{G_{Water}}{G_{Air} + G_{Water}} \quad (D)$$

If we assume $\overline{\omega_2}$ = 0.005 for an initial value, the following values are calculated by means of Eqs. (B), (C), and (D):

$$\Delta\omega_{lm} = 0.0144 - (0.5)(0.005) = 0.01190$$

$$G_{Water} = (3.15 \times 10^{-4})(0.01190) = 3.75 \times 10^{-6}$$

$$\overline{\omega_2}_{cal} = \frac{(3.75 \times 10^{-6})}{(3.75 \times 10^{-6} + 6.41 \times 10^{-4})} = 0.00582$$

A second trial is then carried out by assuming $\overline{\omega_2}$ = 0.00582:

Repeating similar calculations several times, we obtain the following convergence value:

$$G_{Water} = 3.65 \times 10^{-6} \text{ kg s}^{-1}$$

$$\overline{\omega_2} = 0.00565$$

Rate of evaporation = $(3.65 \times 10^{-6})(3600) = 0.013 \text{ kg h}^{-1}$

References

1 H. S. Carslaw and J. C. Jaeger, "*Conduction of Heat in Solids*", 2nd ed., pp. 199–200, Oxford University Press (1959).
2 L. Graetz, "Ueber die Wärmeleitungsfähigkeit von Flüssigkeiten", Annalen der Physik und Chemie, **25**, 337–357 (1885).
3 W. M. Kays and M. E. Crawford, "*Convective Heat and Mass Transfer*", 2nd ed., pp. 106–114, McGraw-Hill (1980).
4 H. Kosuge and K. Asano, "Effect of Column Length and Vapor Condensation on Heat and Mass Transfer in Ternary Distillation by a Wetted-Wall Column", *KAGAKU KOGAKU Ronbunshu*, **10**, [1], 1–6 (1984).
5 H. Kosuge and K. Asano, "Simulation of Performance of Multicomponent Distillation by Sieve Tray Columns", *Sekiyu Gakkaishi (Journal of Petroleum Institute of Japan)*, **28**, [5], 413–416 (1985).
6 J. R. Sellars, M. Tribus, and J. S. Klein, "Heat Transfer to Laminar Flow in a Round Tube or Flat Conduit – The Graetz Problem Extended", *Transactions of the American Society of Mechanical Engineers*, **78**, February, 441–448 (1956).
7 A. H. P. Skelland, "*Diffusional Mass Transfer*", pp. 142–167, Wiley-Interscience (1974).

6
Motion, Heat and Mass Transfer of Particles

6.1
Creeping Flow around a Spherical Particle

The motion of small particles in fluid media is of fundamental concern in various industrial and environmental problems. Figure 6.1 shows the flow around a spherical particle when it is placed in a very slow and uniform flow. Under these circumstances, the stream lines are observed to run almost parallel to the surface of the particle and there is no separation of flow from the surface of the particle. This type of flow is known as *creeping flow*. In a creeping flow, the inertia term in the equation of motion is negligibly small in comparison with the viscosity term and hence can be neglected.

C. G. Stokes presented the following analytical approach to the drag force of a particle in a creeping flow regime [13]. The equation of motion in a creeping flow in terms of the stream functions in spherical coordinates can be expressed as follows:

$$\left[\frac{\partial^2}{\partial r^2} + \frac{\sin\theta}{r^2} \frac{\partial}{\partial \theta}\left(\frac{1}{\sin\theta}\frac{\partial}{\partial \theta}\right)\right]^2 \psi = 0 \tag{6.1}$$

The boundary conditions are:

$r = R = D_P/2$

$$v_r = \frac{1}{r^2 \sin\theta}\frac{\partial \psi}{\partial \theta} = 0 \tag{6.2a}$$

$$v_\theta = \frac{-1}{r \sin\theta}\frac{\partial \psi}{\partial r} = 0 \tag{6.2b}$$

$$r = \infty : v_r = \frac{1}{r^2 \sin\theta}\frac{\partial \psi}{\partial \theta} \to U\cos\theta \tag{6.2c}$$

where D_p is the diameter of the particle [m], $R (= D_p/2)$ is its radius [m], U is the free stream velocity [m s^{-1}], and ψ is the stream function [m^3 s^{-1}].

Mass Transfer. From Fundamentals to Modern Industrial Applications. Koichi Asano
Copyright © 2006 WILEY-VCH Verlag GmbH & Co. KGaA, Weinheim
ISBN: 3-527-31460-1

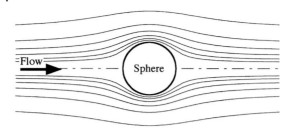

Fig. 6.1 Creeping flow around a spherical particle.

If we assume the following equation for the stream function, which will automatically satisfy the boundary conditions at a point far away from the particle, Eq. (6.2.c):

$$\psi \equiv f(r)\sin^2\theta \tag{6.3}$$

the analytical solution of Eq. (6.1) can easily be obtained by the method of separation of variables. Substituting Eq. (6.3) into Eq. (6.1) and integrating the resultant equation with respect to r by considering the boundary conditions, we obtain the following equation for the stream function in a creeping flow:

$$\psi = \left\{1 - \frac{3}{2}\left(\frac{R}{r}\right) + \frac{1}{2}\left(\frac{R}{r}\right)^3\right\}\left(\frac{U}{2}\right)r^2\sin^2\theta \tag{6.4}$$

Substituting Eq. (6.4) into Eqs. (6.2.a) and (6.2.b), the radial and tangential components of the velocity, v_r and v_θ, are given by the following equations:

$$\frac{v_r}{U} = \left\{1 - \frac{3}{2}\left(\frac{R}{r}\right) + \frac{1}{2}\left(\frac{R}{r}\right)^3\right\}\cos\theta \tag{6.5a}$$

$$\frac{v_\theta}{U} = -\left\{1 - \frac{3}{4}\left(\frac{R}{r}\right) - \frac{1}{4}\left(\frac{R}{r}\right)^3\right\}\sin\theta \tag{6.5b}$$

The shear stress due to the friction on the surface of the particle at angle θ is given by the following equation:

$$\tau_{r\theta}|_{r=R} = -\mu\left(\frac{\partial v_\theta}{\partial r}\right)_{r=R} = \frac{3\mu U}{2R}\sin\theta \tag{6.6}$$

The *frictional drag* acting on the surface of the particle, F_f [N], is calculated by integrating the axial component of the shear stress on the surface, $\tau_{r\theta}|_{r=R}$ [N m^{-2}], with a surface element of $2\pi R^2\sin\theta\,d\theta$, as:

$$F_f = 2\pi \int_0^\pi (\tau_{r\theta}|_{r=R} \sin\theta) R^2 \sin\theta \, d\theta$$

$$= 2\pi \int_0^\pi \left(\frac{3\mu U}{2R} \sin^2\theta\right) R^2 \sin\theta \, d\theta = 2\pi\mu D_p U \tag{6.7}$$

On the other hand, the pressure distribution on the surface of the particle is given by substituting Eqs. (6.5.a) and (6.5.b) into the equation of motion as:

$$P = P_\infty - \frac{3\mu U}{2R}\left(\frac{R}{r}\right)^2 \cos\theta \tag{6.8}$$

The *form drag* (or pressure drag), F_p [N], the drag due to the axial component of the pressure on the surface of the particle, can be calculated by integrating the axial component of the normal stress on the surface of the particle, $P|_{r=R}\cos\theta$, with a small surface element, $2\pi R^2 \sin\theta \, d\theta$, as follows:

$$F_P = 2\pi \int_0^\pi \left(\frac{3\mu U}{2R}\cos^2\theta\right) R^2 \sin\theta \, d\theta = \pi\mu D_p U \tag{6.9}$$

The *drag force* of the particle, F_D [N], is calculated as the sum of the frictional drag, F_f and the form drag, F_p:

$$F_D = F_f + F_P = 3\pi\mu D_p U \tag{6.10}$$

Equation (6.10) is referred to as *Stokes' law of resistance*, and when it is applicable the flow around the particle is called *Stokes' flow*.

If we define the drag coefficient of the spherical particle by the following equation:

$$C_D \equiv \frac{F_D}{\left(\frac{\pi D_p^2}{4}\right)\left(\frac{\rho U^2}{2}\right)} \tag{6.11}$$

then *Stokes' law of resistance* can also be described in terms of the drag coefficients and the Reynolds numbers by the following equation:

$$C_D = 24/Re_p \tag{6.12}$$

where Re_p is the *particle Reynolds number* defined by the following equation:

$$Re_p \equiv \rho D_p U/\mu \tag{6.13}$$

and ρ is the density of the fluid [kg m^{-3}] and μ its viscosity [Pa s]. An important fact about *Stokes' law* is that it is based on the assumption of creeping flow and consequently is only valid for very low Reynolds numbers:

$$Re_p \leq 0.1$$

6.2
Motion of Spherical Particles in a Fluid

6.2.1
Numerical Solution of the Drag Coefficients of a Spherical Particle in the Intermediate Reynolds Number Range

If the Reynolds number becomes moderate or large, the relative order of magnitude of the inertia term in the equation of motion becomes considerable and we can no longer neglect it. Under these circumstances, the assumption of creeping flow no longer holds and the nonlinear terms in the equation of motion cannot be neglected. Thus, obtaining an analytical solution, as in the case of *the Stokes' flow* described in the previous section, becomes practically impossible and only a numerical approach is possible.

The equation of motion in terms of the dimensionless stream functions in spherical coordinates can be written as follows:

$$\sin\theta \frac{\partial \psi'}{\partial r'} \frac{\partial}{\partial \theta}\left(\frac{E^2\psi'}{r'^2 \sin^2\theta}\right) - \sin\theta \frac{\partial \psi'}{\partial \theta} \frac{\partial}{\partial r'}\left(\frac{E^2\psi'}{r'^2 \sin^2\theta}\right) = \frac{2}{Re_p} E^4 \psi' \quad (6.14)$$

$$E^2 = \left(\frac{\partial^2}{\partial r'^2} + \frac{\sin\theta}{r'^2}\frac{\partial}{\partial \theta}\left(\frac{1}{\sin\theta}\frac{\partial}{\partial \theta}\right)\right) \quad (6.15)$$

The boundary conditions are:

$$\theta = 0: \quad \psi' = 0 \quad (6.16a)$$

$$\theta = \pi: \quad \psi' = 0 \quad (6.16b)$$

$$r' = 1: \quad \psi' = 0 \quad (6.16c)$$

$$r' = r'_\infty: \quad \psi' = \frac{r'^2_\infty}{2} \sin^2\theta \quad (6.16d)$$

where r' and ψ' are dimensionless variables defined by the following equations:

$$r' = r/R \quad (6.17a)$$

$$\psi' = \psi/R^2 U \quad (6.17b)$$

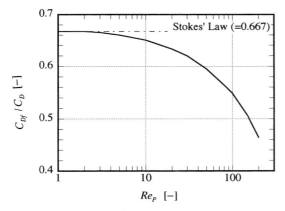

Fig. 6.2 Variation in the relative order of magnitude of the frictional drag compared to the total drag of a spherical particle with Reynolds number [2].

Detailed descriptions of the numerical calculations are very complicated and are thus beyond the scope of this book. We will only describe the final results of the calculations.

A. E. Hamielec et al. [7] presented a numerical approach to this problem for a Reynolds number range up to 100. B. P. LeClair et al. [10] describe a similar approach for a Reynolds number range up to 400. A more extensive treatment, including the effect of mass injection, was presented by P. Chuchottaworn, A. Fujinami, and K. Asano [3].

Figure 6.2 shows the relative order of magnitude of the viscous drag compared to the total drag of the spherical particle, as calculated by Chuchottaworn [2]. The figure indicates that the relative order of magnitude of the viscous drag decreases gradually from the theoretical value given by *Stokes' law of resistance* ($= 2/3$) as the Reynolds number increases.

6.2.2
Correlation of the Drag Coefficients of a Spherical Particle

A numerical approach to the drag coefficients of spherical particles can be made up to a Reynolds number range of several hundreds, but even this range is not wide enough to cover the necessary Reynolds number range needed for practical use. Consequently, correlations of the drag coefficients applicable for a wide range of Reynolds numbers based on sufficiently reliable experimental data are needed.

C. E. Lapple and C. B. Shepherd [9] first proposed their famous correlation for the drag coefficients of a spherical particle based on a wide range of experimental data. Many correlations have since been reported, which are well summarized by Clift et al. [6], among which is the useful correlation by H. Brauer and D. Sucker [1].

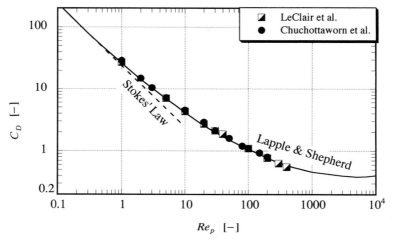

Fig. 6.3 Correlation of drag coefficients of a spherical particle; comparison of Lapple–Shepherd's correlation with numerical data obtained by Chuchottaworn et al. [3] and LeClair et al. [10].

$$C_D = \frac{24}{Re_p} + \frac{3.73}{Re_p^{1/2}} - \frac{4.83 \times 10^{-3} Re_p^{1/2}}{1 + 3 \times 10^{-6} Re_p^{3/2}} + 0.49 \quad (6.18)$$

$$0 < Re_p < 3.5 \times 10^5$$

Figure 6.3 shows a comparison of the correlation of drag coefficients of spherical particles with the numerical data obtained by B. P. LeClair et al. [10] and that obtained by P. Chuchottaworn, A. Fujinami, and K. Asano [3]. The solid line in the figure represents the Lapple–Shepherd correlation [9], which may be adequately approximated by the following equation for the Reynolds number range $Re_p < 1000$.

$$C_D = \frac{24}{Re_p}(1 + 0.125 \, Re_p^{0.72}) \quad (6.19)$$

6.2.3
Terminal Velocity of a Particle

The motion of spherical particles moving in a fluid medium under the influence of gravity depends on the force balance between gravity, the buoyancy force, and the drag force. The equation of motion for a particle moving in a fluid can be written as follows:

$$\frac{dv}{dt} = \left(1 - \frac{\rho_f}{\rho_p}\right)g - \frac{3}{4}\frac{C_D \rho_f}{D_p \rho_p}v^2 \quad (6.20)$$

where C_D is the drag coefficient [–], D_p is the diameter of the particle [m], g is the acceleration due to gravity [m² s⁻¹], v is the velocity of the particle [m s⁻¹], ρ_p is the density of the particle [kg m⁻³], and ρ_f is the density of the fluid [kg m⁻³].

The first term on the right-hand side of the above equation represents the acceleration due to gravity and the buoyancy force, while the second term represents the deceleration due to the drag force. In the initial stage of motion, the particle is accelerated by the effect of the first term and is less affected by the second term. The velocity of the particle therefore increases with time. However, as the velocity of the particle increases, the drag force, which is proportional to the square of the velocity, increases much faster than the acceleration term, and soon the two terms balance one another. After a certain time has elapsed, the motion of the particle approaches a steady state. The velocity of the particle in this steady motion is called the *terminal velocity*.

Since the unsteady term in Eq. (6.20) is zero at the terminal velocity:

$$dv/dt = 0$$

we have the following equation for the terminal velocity from Eq. (6.20):

$$v_\infty = \left(\frac{4(\rho_p - \rho_f)D_p g}{3\rho_f C_D}\right)^{1/2} \qquad (6.21)$$

If the free stream velocity is very low and the flow around the particle is a creeping flow, then Stokes' law, Eq. (6.12), is applicable. In this special case, Eq. (6.21) reduces to the following equation:

$$v_\infty = \frac{(\rho_p - \rho_f)D_p^2 g}{18\mu_f} \qquad (6.22)$$

where μ_f is the viscosity of the fluid [Pa s].

Example 6.1

A water drop of diameter 1 mm is falling freely in air at a temperature of 20 °C and a pressure of 1 atm under the influence of gravity. Calculate the terminal velocity of the water drop, if the variation in its diameter is negligibly small. Calculate the distance that the drop has to travel to reach its terminal velocity, assuming that its initial velocity is zero ($v_0 = 0$).

Solution

The physical properties under the given conditions are as follows:

$$\rho_p = 1000 \text{ kg m}^{-3}, \rho_f = 1.20 \text{ kg m}^{-3}, \mu_f = 1.81 \times 10^{-5} \text{ Pa s}$$

In this case, we cannot apply *Stokes' law* because the drop diameter is too large and the Reynolds number for the drop is not within the range for *Stokes' flow*. We thus apply Eq. (6.21) and carry out the calculation by a *trial and error* method.

The first trial:
For an initial calculation, we assume the following value:

$v_0 = 3.0 \text{ m s}^{-1}$

$Re_p = (1.2)(1.0 \times 10^{-3})(3.0)/(1.81 \times 10^{-5}) = 198.8$

Substituting the above value into Eq. (6.19), we obtain the following value:

$$C_D = \frac{24}{198.8}\{1 + 0.125(198.8)^{0.72}\} = 0.80$$

Substituting the above value into Eq. (6.21), we have:

$$v_1 = \left\{\frac{(4)(1000 - 1.20)(1.0 \times 10^{-3})(9.81)}{(3)(1.20)(0.80)}\right\}^{1/2} = 3.68 \text{ m s}^{-1}$$

The second trial:
Substituting the above value, we have the following results after the second trial:

$v_1 = 3.68 \text{ m s}^{-1}$
$Re_p = 244.2$
$C_D = 0.74$
$v_2 = 3.83 \text{ m s}^{-1}$

Repeating similar calculations several times, we obtain a converged value of the terminal velocity of:

$v = 3.87 \text{ m s}^{-1}$

The distance that the drop has to travel to reach its terminal velocity can be calculated by numerical integration of Eq. (6.20) under the boundary condition of $v_0 = 0$.
The calculation is carried out by using the following equations until convergence is obtained:

$$\left(\frac{dv}{dt}\right)_i = \left(1 - \frac{\rho_f}{\rho_p}\right)g - \frac{3}{4}\frac{C_D \rho_f}{D_p \rho_p}v_i^2 \quad \text{(A)}$$

$$v_{i+1} = v_i + \left(\frac{dv}{dt}\right)_i \Delta t \quad \text{(B)}$$

$$y_{i+1} = y_i + (v_i + v_{i+1})(1/2)\Delta t \quad \text{(C)}$$

Table 6.1 summarizes the results of the numerical integration. They indicate that the drop will reach a terminal velocity of $v_\infty = 3.87 \text{ m s}^{-1}$ at a distance of 6 m from the starting point. Figure 6.4 shows the motion of a water drop falling freely in air.

Table 6.1 Free fall of a water drop (D_{p0} = 1.0 mm) (Example 6.1).

t [s]	v [m/s]	y [m]	Re_p	C_D	dv/dt
0.00	0.00	0.00	0.0	0.000	9.798
0.10	0.98	0.05	65.0	1.302	8.673
0.20	1.85	0.19	122.5	0.977	6.799
0.30	2.53	0.41	167.5	0.858	4.865
0.40	3.01	0.69	199.8	0.801	3.253
0.50	3.34	1.00	221.4	0.770	2.074
0.60	3.55	1.35	235.1	0.752	1.282
0.70	3.67	1.71	243.6	0.742	0.776
0.80	3.75	2.08	248.8	0.737	0.465
0.90	3.80	2.46	251.8	0.733	0.276
1.00	3.83	2.84	253.7	0.731	0.163
1.10	3.84	3.22	254.7	0.730	0.096
1.20	3.85	3.61	255.4	0.729	0.057
1.30	3.86	3.99	255.8	0.729	0.033
1.40	3.86	4.38	256.0	0.729	0.020
1.50	3.86	4.76	256.1	0.729	0.012
1.60	3.86	5.15	256.2	0.729	0.007
1.70	3.86	5.54	256.2	0.729	0.004
1.80	3.87	5.92	256.3	0.729	0.002
1.90	3.87	6.31	256.3	0.729	0.001

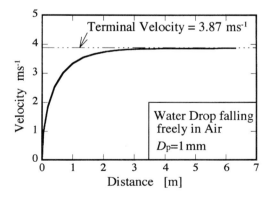

Fig. 6.4 Motion of a water drop falling freely in air.

6.3
Heat and Mass Transfer of Spherical Particles in a Stationary Fluid

Equation (6.22) indicates that the terminal velocity of a spherical particle in Stokes' flow is proportional to the square of its diameter. This means that the relative velocity of the particle compared to that of the fluid is almost zero and that the particle is floating motionless in the fluid. We will consider heat and mass transfer in this special case.

The energy equation for a very small particle in a stationary fluid can be written as follows:

$$\frac{\partial}{\partial r}\left(r^2 \frac{\partial T}{\partial r}\right) = 0 \tag{6.23}$$

The boundary conditions are:

$$r = R = D_p/2 : T = T_s, \tag{6.24a}$$

$$r = \infty : T = T_\infty \tag{6.24b}$$

Integrating Eq. (6.23) by considering the boundary conditions, we obtain the following equation:

$$\frac{T - T_\infty}{T_s - T_\infty} = \frac{R}{r} \tag{6.25}$$

The local heat flux on the surface of the particle, q [W m^{-2}], is given by:

$$q = -\kappa \left(\frac{dT}{dr}\right)_{r=R} = \frac{\kappa (T_s - T_\infty)}{R} \tag{6.26}$$

Rearranging Eq. (6.26) into dimensionless form, we obtain the following theoretical equation for the Nusselt number:

$$Nu_p = \frac{q D_p}{\kappa (T_s - T_\infty)} = 2 \tag{6.27}$$

A similar equation is obtained for the mass transfer of a very small particle in a stationary fluid, that is:

$$Sh_p(1 - \omega_s) = \frac{N_A (1 - \omega_s) D_p}{D(\omega_s - \omega_\infty)} = 2 \tag{6.28}$$

where unidirectional diffusion is assumed.

6.4
Heat and Mass Transfer of Spherical Particles in a Flow Field

6.4.1
Numerical Approach to Mass Transfer of a Spherical Particle in a Laminar Flow

The mass transfer of a spherical particle in a laminar flow of intermediate Reynolds number range can be obtained through numerical solution of the diffusion equation in terms of the dimensionless stream functions in spherical coordinates, together with the equation of motion, Eq. (6.14). The detailed calculations are very complicated and beyond the scope of this book, and so for the convenience of the reader we will only show the governing equation and the boundary conditions:

Diffusion equation:

$$\frac{Re_p Sc}{2}\left(\frac{\partial \psi'}{\partial r'}\frac{\partial \theta_c}{\partial \theta} - \frac{\partial \psi'}{\partial \theta}\frac{\partial \theta_c}{\partial r'}\right) = \sin\theta \left(r'^2 \frac{\partial^2 \theta_c}{\partial r'^2} + 2r' \frac{\partial \theta_c}{\partial r'} + \cot\theta \frac{\partial \theta_c}{\partial \theta} + \frac{\partial^2 \theta_c}{\partial \theta^2}\right) \tag{6.29}$$

where θ_c is the dimensionless concentration defined by the following equation:

$$\theta_c = \frac{\omega_s - \omega}{\omega_s - \omega_\infty} \tag{6.30}$$

The boundary conditions are:

$\theta = 0: \quad \psi' = 0, \quad\quad \partial\theta_c/\partial\theta = 0$ (6.31a)

$\theta = \pi: \quad \psi' = 0, \quad\quad \partial\theta_c/\partial\theta = 0$ (6.31b)

$r' = 1: \quad \psi' = 0, \quad\quad \theta_c = 0$ (6.31c)

$r' = r'_\infty: \quad \psi = (r'^2_\infty/2)\sin^2\theta, \quad \theta_c = 1$ (6.31d)

An equation similar to Eq. (6.29) can also be obtained for the heat transfer of a spherical particle in a laminar flow by replacing Sc with Pr and θ_c with the dimensionless temperature $\theta_T (\equiv T_s - T)/(T_s - T_\infty)$.

S. E. Woo and A. E. Hamielec [16] described a numerical approach to this problem for the case of $Pr = 0.71$. Later, P. Chuchottaworn, A. Fujinami, and K. Asano [4] presented a similar approach to the problem for the wider range of Schmidt numbers of 0.5 to 2.0. Figure 6.5 shows the distribution of the local Nusselt numbers, Nu_θ, and the dimensionless local diffusional fluxes over the surface of a particle, $Sh_\theta(1 - \omega_s)$, which are defined by the following equations:

$$Nu_\theta \equiv q_\theta D_p / \{\kappa(T_s - T_\infty)\} \tag{6.32a}$$

$$Sh_\theta(1 - \omega_s) \equiv J_{A\theta} D_p / \{\rho D(\omega_s - \omega_\infty)\} \tag{6.32b}$$

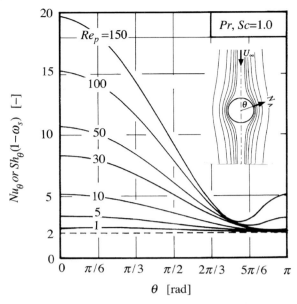

Fig. 6.5 Distribution of the local heat and diffusional fluxes on the surface of a spherical particle [4].

The figure indicates that the local diffusional flux (or heat flux) attains its maximum value at the *forward stagnation point* and decreases as the angle θ increases, and that the distribution becomes more flat and finally approaches the theoretical value for the stationary fluid as the Reynolds number decreases.

$$Nu_\theta = Sh_\theta(1 - \omega) = 2$$

The slight increase in the local flux to the rear of the particle in the case of higher Reynolds numbers (Re_p = 100, 150) may be attributed to the effect of small eddies in the wake region.

6.4.2
The Ranz–Marshall Correlation and Comparison with Numerical Data

Although the numerical approach to the heat and mass transfer of a spherical particle can be carried out up to Reynolds numbers of several hundreds, this range is not sufficient for practical use. A reliable correlation applicable over a wide range of Reynolds numbers is needed.

W. E. Ranz and W. R. Marshall [11] proposed the following well-known correlation for the heat and mass transfer of a spherical particle on the basis of a wide variety of experimental data on the evaporation of liquid droplets in air.

6.4 Heat and Mass Transfer of Spherical Particles in a Flow Field

$$1 < Re_p Pr^{2/3} < 5 \times 10^4$$

$$Nu_p = 2 + 0.6 Re_p^{1/2} Pr^{1/3} \tag{6.33}$$

$$1 < Re_p Sc^{2/3} < 5 \times 10^4$$

$$Sh_p (1 - \omega_s) = 2 + 0.6\, Re_p^{1/2} Sc^{1/3} \tag{6.34}$$

The basic idea of their correlation is that in the low Reynolds number regime the correlation must approach asymptotically the theoretical value for the stationary fluid, whereas in the large Reynolds number regime the flow around the sphere is turbulent and should follow *Chilton and Colburn's analogy*. Although the correlation is a simple combination of the two asymptotic equations for the extreme cases, that is, stationary fluid and very large free stream velocity, it can be applied for a wide range of Reynolds numbers with fairly good accuracy. For this reason, the correlation is widely used in practical applications.

Figure 6.6 shows a comparison of the Ranz–Marshall correlation with numerical data obtained by Woo and Hamielec [15] and by Chuchottaworn et al. [4].

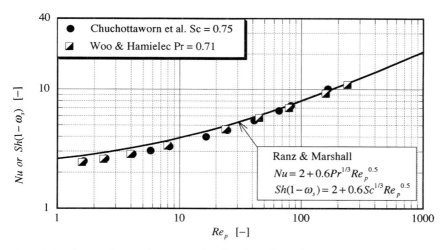

Fig. 6.6 Correlation of heat and mass transfer of a spherical particle; comparison of Ranz and Marshall's correlation with numerical data obtained by Chuchottaworn et al. [4] and by Woo and Hamielec [16].

Although the Ranz–Marshall correlation gives fairly good agreement with a wide range of observed data, it shows about 10% higher values than the numerical data in the Reynolds number range $Re_p < 200$. This may be attributed to the asymptotic nature of the correlation. Chuchottaworn et al. [4] recommended the following correlations for this practically important Reynolds number range on the basis of their numerical data.

6 Motion, Heat and Mass Transfer of Particles

$1 \leq Re_p \leq 200, \; 0.5 \leq Pr \leq 2.0$:

$$Nu_p = 2 + 0.37 Re_p^{0.61} Pr^{0.51} \qquad (6.35)$$

$1 \leq Re_p \leq 200, \; 0.5 \leq Sc \leq 2.0$:

$$Sh_p(1 - \omega_s) = 2 + 0.37 Re_p^{0.61} Sc^{0.51} \qquad (6.36)$$

Example 6.2
A water drop of initial diameter 1 mm is injected downwards from a nozzle into air at 1 atm and 317 K. Calculate the diameter of the drop when it reaches a distance of 5 m from the nozzle if its initial temperature is 293 K and its initial velocity is 3 m s^{-1}.

Solution
If we tentatively assume that the surface of the drop is at wet-bulb temperature, and that this is 288 K, the physical properties of the system are as follows:

$\omega_s = 0.0103 \; [-]$
$\rho = 1.23 \; \text{kg m}^{-3}$
$D = 2.40 \times 10^{-5} \; \text{m}^2 \; \text{s}^{-1}$
$\mu = 1.92 \times 10^{-5} \; \text{Pa s}$

Example 6.1 indicates that the terminal velocity of the water drop is $v_\infty = 3.87$ m s^{-1}, which is close to the initial velocity. Thus, we will assume that the drop soon attains its terminal velocity.

$$Sc = \frac{(1.92 \times 10^{-5})}{(1.23)(2.40 \times 10^{-5})} = 0.65$$

$$Re_p = \frac{(1.23)(1.0 \times 10^{-3})(3.87)}{(1.92 \times 10^{-5})} = 248$$

$$Sh(1 - \omega_s) = 2 + 0.6(0.65)^{1/3}(248)^{0.5} = 10.2$$

$$N_A = \frac{(10.2)(1.23)(2.40 \times 10^{-5})}{(1.0 \times 10^{-3})} \left(\frac{0.0103}{1 - 0.0103}\right)$$

$$= 3.13 \times 10^{-3} \; \text{kg m}^{-2} \; \text{s}^{-1}$$

Surface area of the drop $= (3.14)(1.0 \times 10^{-3})^2 = 3.14 \times 10^{-6} \; \text{m}^2$

Rate of evaporation $= (3.14 \times 10^{-6})(3.13 \times 10^{-3})(5/3.87) = 1.27 \times 10^{-8}$ kg

Initial mass of the drop $= (3.14/6)(1.0 \times 10^{-3})^3(1000) = 5.23 \times 10^{-7}$ kg

$$\left(\frac{D_p}{D_{p0}}\right) = \left\{1 - \frac{(1.27 \times 10^{-8})}{(5.23 \times 10^{-7})}\right\}^{1/3} = 0.99$$

6.4.3
Liquid-Phase Mass Transfer of a Spherical Particle in Stokes' Flow

The orders of magnitude of Schmidt numbers for a liquid range from several hundreds to twenty to thirty thousands. Under such circumstances, the concentration boundary layer lies very near to the surface of the particle in comparison with the velocity boundary layer, as discussed in Section 3.4. V. G. Levich [11] discussed the liquid-phase mass transfer of a particle in Stokes' flow by considering this fact.

The diffusion equation for Stokes' flow in the liquid phase can be approximated by the following equation:

$$\frac{v_\theta}{R} \frac{\partial \omega}{\partial \theta} = D \frac{\partial^2 \omega}{\partial y^2} \tag{6.37}$$

The stream function for very large Schmidt numbers can be approximated by the following equation:

$$v_\theta = -\frac{1}{r \sin\theta} \frac{\partial \psi}{\partial r} \approx \frac{-3}{2} \left(\frac{r}{R} - 1\right) U \sin\theta \tag{6.38}$$

Substituting Eq. (6.38) into Eq. (6.37) and solving the resultant equation, we obtain the following equation:

$$Sh_p (1 - \omega_s) = 0.992 Re_p^{1/3} Sc^{1/3} \tag{6.39}$$

Because of the approximate nature of the derivation of the above equation, it cannot be applied in the following range:

$$Pe_M \equiv Re_p Sc \leq 8.2$$

In this range, it gives much lower values than the theoretical value for the stationary fluid ($Sh_p (1 - \omega_s) = 2$).

6.5
Drag Coefficients, Heat and Mass Transfer of a Spheroidal Particle

In practical applications, particles are not always spherical in shape but in most cases are deformed to a certain degree. Small-sized droplets in the gas phase or bubbles in the liquid phase are usually spherical in shape, but medium-sized drops or bubbles are usually deformed to an oblate spheroid shape, as is formed by rotation of an ellipse about its short axis.

P. Chuchottaworn and K. Asano [5] presented a numerical approach to this problem by using the finite difference method with spheroidal coordinate sys-

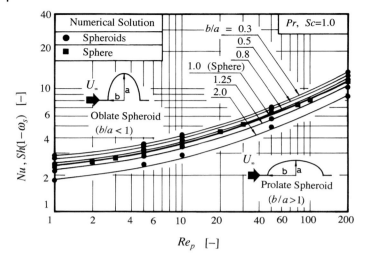

Fig. 6.7 Effect of aspect ratio on heat and mass transfer in a spheroidal particle [5].

tems. Figure 6.7 shows heat and mass transfer of spheroidal particles. In the figure, the characteristic length of the particle is tentatively taken as the maximum diameter normal to the direction of flow, a [m], and the following dimensionless numbers are used in the correlation:

$$Nu \equiv qa/\kappa (T_s - T_\infty) \tag{6.40}$$

$$Sh(1 - \omega_s) \equiv J_A a/\rho D (\omega_s - \omega_\infty) \tag{6.41}$$

$$Re \equiv \rho a U/\mu \tag{6.42}$$

The bold solid line in the figure represents the correlation for spherical particles. The fact that a systematic deviation from the standard correlation for the sphere with the aspect ratio ($E \equiv b/a$) is observed suggests the following correlation for heat and mass transfer of spheroidal particles:

$$\frac{Nu}{Nu_p} = \frac{Sh(1 - \omega_s)}{Sh_p(1 - \omega_s)} = \exp(-0.28(E - 1)) \tag{6.43}$$

The ranges of the variables are as follows:

$1 \leq Re \leq 200$

$0.5 \leq Pr, \ Sc \leq 2.0$

$0.3 \leq E \leq 2.0$

where Nu_p and $Sh_p(1-\omega_s)$ are the Nusselt number and the dimensionless diffusional flux for a spherical particle, respectively.

In the same way as the characteristic length of the particles, the volume average diameter defined by the following equation:

$$D_v \equiv \left(\frac{6V}{\pi}\right)^{1/3} = (a^2 b)^{1/3} \tag{6.44}$$

is sometimes used in the literature. The possibility of using the volume mean diameter in the same way as a characteristic length has also been tested, but no systematic relationship was obtained.

6.6
Heat and Mass Transfer in a Fluidized Bed

6.6.1
Void Function

Heat and mass transfer in fluidized beds is one of the important concerns of chemical engineers. In describing heat and mass transfer in a fluidized bed, the *void function* defined by the following equation is used:

(heat and mass transfer in fluidized bed)
$$= \text{(void function)(heat and mass transfer of single particle)} \tag{6.45}$$

Obtaining reliable correlations for the void function is therefore one of the important concerns in the study of fluidized beds. Due to the complicated nature of transport phenomena in fluidized beds, a theoretical approach to this problem is practically impossible and only experimental approaches can be implemented. Unfortunately, existing data on the void function are widely scattered and no reliable correlation has ever been obtained. The main reason for this may stem from the fact that most studies have been based on measurements of the drying of wet particles in fluidized beds, where insufficient distinction is made between true mass transfer and pseudo mass transfer during the decreasing drying rate period.

6.6.2
Interaction of Two Spherical Particles of the Same Size in a Coaxial Arrangement

I. Taniguchi and K. Asano [14] described a numerical approach to delineating the effect on drag coefficients and on heat and mass transfer of interactions of spherical particles of the same size arranged coaxially by using the finite difference method with bipolar coordinates. Figure 6.8 shows the effect on heat and mass transfer of the interaction between two spherical particles of the same size in a coaxial arrangement. The upstream particle is little affected by the presence of the

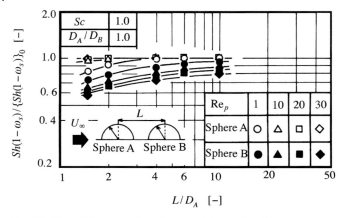

Fig. 6.8 Effect of the interaction of two coaxial particles of the same size on the rate of mass transfer [14].

downstream particle, but the downstream particle is significantly affected by the presence of the upstream particle, and the effect becomes considerable as the distance between the two particles decreases and the Reynolds number increases. This may be due to the effect of the *wake* of the upstream particle, which will greatly reduce the local mass or heat flux in the region near the forward stagnation point of the downstream particle, where the local flux would be highest if the upstream particle were to be eliminated.

Numerical results relating to the effect of interactions on the mass transfer of the downstream particle are well correlated by the following equation:

$$\frac{Sh(1-\omega_s)}{Sh_0(1-\omega_s)} = \frac{1}{1.0 + 0.77(L/D_p)^{-n}} \tag{6.46}$$

$$n = -1.38 Re_p^{-0.26} \tag{6.47}$$

where D_p is the diameter of the particles [m], and L is the distance between the centers of the two particles [m].

6.6.3
Simulation of the Void Function

I. Taniguchi and K. Asano [15] described a simulation approach to the prediction of the void function of a fluidized bed on the basis of the following assumptions:
1) Numerical correlation of the effect of the interaction of two coaxially arranged particles of the same size; Eq. (6.46).
2) Empirical correlation of the interaction of two particles of the same size in an eccentric arrangement.

6.6 Heat and Mass Transfer in a Fluidized Bed

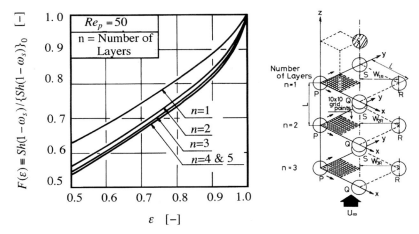

Fig. 6.9 Simulation of void functions for heat and mass transfer in a fluidized bed based on a multilayer particle model [15].

They further assumed that the layers of particles of the same size are in a rectangular arrangement, with the interlayer distance being equal to the distance between the particles, and that these layers move randomly in the horizontal direction as shown in the right-hand part of Fig. 6.9.

Figure 6.9 shows the results of their simulation of the void function, which indicate that the effect of the interaction of the particle layers converges to a single line if the number of layers exceeds five, and the following equation is obtained for the void function:

$$F(\varepsilon) \equiv \frac{\{Sh_p(1-\omega_s)\}_{\text{Fluidized Bed}}}{\{Sh_p(1-\omega_s)\}_0} = \frac{1}{1+1.32(1-\varepsilon)^{0.76}} \qquad (6.47)$$

where $\{Sh_p(1-\omega_s)\}_0$ is the diffusional flux of an individual spherical particle estimated by the Ranz–Marshall correlation and ε is the voidage of the fluidized bed. Equation (6.47) shows good agreement with experimental data for the sublimation of naphthalene in air in a fluidized bed as obtained by T. H. Hsiung and G. Thodos [8]. Heat and mass transfer in the fluidized bed may be adequately predicted by the use of Eq. (6.47) together with the Ranz–Marshall correlation.

References

1 H. Brauer and D. Sucker, "Flow about Plates, Cylinders and Spheres", *International Chemical Engineering*, **18**, [3], 367–374 (1978); *Chemie-Ingenieur-Technik*, **48**, [8], 665–671 (1976).

2 P. Chuchottaworn, "*Drag Coefficients, Heat and Mass Transfer of an Evaporating Liquid Drop*", Ph.D. Thesis in Chemical Engineering, Tokyo Institute of Technology, (1984).

3 P. Chuchottaworn, A. Fujinami, and K. Asano, "Numerical Analysis of the Effect of Mass Injection and Suction on Drag Coefficients of a Sphere", *Journal of Chemical Engineering of Japan*, **16**, [1], 18–24 (1983).

4 P. Chuchottaworn, A. Fujinami, and K. Asano, "Numerical Analysis of Heat and Mass Transfer from a Sphere with Surface Mass Injection and Suction", *Journal of Chemical Engineering of Japan*, **17**, [1], 1–7 (1984).

5 P. Chuchottaworn and K. Asano, "Numerical Analysis of Drag Coefficients and Heat and Mass Transfer of Spheroidal Drops", *Journal of Chemical Engineering of Japan*, **19**, [3], 208–214 (1986).

6 R. Clift, J. R. Grace, and M. E. Weber, "*Bubbles, Drops, and Particles*", pp. 111–112, Academic Press (1978).

7 A. E. Hamielec, T. W. Hoffman, and L. L. Ross, "Numerical Solution of the Navier–Stokes Equation for Flow Past Spheres", *A. I. Ch. E. Journal*, **13**, [2], 212–219 (1967).

8 T. H. Hsiung and G. Thodos, "Mass Transfer in Gas-Fluidized Beds: Measurement of Actual Driving Forces", *Chemical Engineering Science*, **32**, 581–592 (1977).

9 C. E. Lapple and C. B. Shepherd, "Calculation of Particle Trajectories", *Industrial Engineering Chemistry*, **32**, [5], 605–617 (1940).

10 B. P. LeClair, A. E. Hamielec, and H. R. Pruppacher, "A Numerical Study of the Drag on a Sphere at Low and Intermediate Reynolds Numbers", *Journal of the Atmospheric Sciences*, **27**, [3], 308–315 (1970).

11 V. G. Levich, "*Physicochemical Hydrodynamics*", pp. 80–87, Prentice-Hall (1960).

12 W. E. Ranz and W. R. Marshall, "Evaporation from Drops, Part I and Part II", *Chemical Engineering Progress*, **48**, [3], 141–146, [4], 173–180 (1952).

13 J. C. Slattery, "Momentum, Energy and Mass Transfer in Continua", pp. 110–114, McGraw-Hill (1972).

14 I. Taniguchi and K. Asano, "Numerical Analysis of Drag Coefficients and Mass Transfer of Two Adjacent Spheres", *Journal of Chemical Engineering of Japan*, **20**, [3], 287–294 (1987).

15 I. Taniguchi and K. Asano, "Experimental Study of the Effect of Neighboring Solid Spheres on the Rates of Evaporation of a Drop and Prediction of Void Function", *Journal of Chemical Engineering of Japan*, **25**, [3], 321–326 (1992).

16 S. E. Woo and A. E. Hamielec, "A Numerical Method of Determining the Rate of Evaporation of Small Water Drop Falling at Terminal Velocity in Air", *Journal of the Atmospheric Sciences*, **28**, [11], 1448–1454 (1971).

7
Mass Transfer of Drops and Bubbles

7.1
Shapes of Bubbles and Drops

A characteristic feature of the mass transfer of bubbles and drops is that they have free surfaces determined by the force balance acting thereon. The motion and mass transfer of bubbles and drops are closely related to the shapes of their surfaces, and the latter are quite sensitive to the conditions of flow as well as to the physical properties of the systems. In this regard, they are quite different from those of solid particles. For small bubbles or droplets the surface tension is dominant at the interface, and as a result they tend to be spherical in shape. For intermediate sized bubbles and drops, however, the shapes are deformed and sometimes becomes unstable.

Figure 7.1 shows schematic pictures of bubbles in the liquid phase. Small bubbles of diameter < 1 mm are usually spherical in shape, whereas bubbles of intermediate size tend to deform from spherical to oblate spheroid, with the minor axis of the ellipse as the axis of rotation, as the bubble size increases. Larger bubbles are even more deformed to mushroom-like shapes, and their shapes are quite unstable and may even be oscillating. In an ascending motion, small bubbles move straight upwards, medium-sized bubbles ascend in a spiral motion, while large bubbles tend to oscillate in an irregular manner during their ascent.

Drops in the liquid phase of diameter < 1 mm are spherical, whereas medium-sized drops of several millimeters in diameter tend to deform from spherical to oblate spheroid.

Drops in the gas phase are rather stable in comparison with those in the liquid phase, and are nearly spherical in shape if their sizes are less than a few millimeters in diameter.

a. Sphere   b. Oblate Spheroid   c. Spherical Cap

Fig. 7.1 Shapes of bubbles.

Mass Transfer. From Fundamentals to Modern Industrial Applications. Koichi Asano
Copyright © 2006 WILEY-VCH Verlag GmbH & Co. KGaA, Weinheim
ISBN: 3-527-31460-1

Bubbles and drops are referred to as the *dispersed phase*, while the fluid containing the dispersed phase is referred to as the *continuous phase*. Since heat and mass transfer in the dispersed phase and those in the continuous phase are affected by quite different factors, and mass transfer resistances of bubbles and drops exist on both sides of the interface, we must be careful in deciding on which side the resistance constitutes the rate-controlling process. For example, the evaporation of drops in the gas phase is continuous-phase mass transfer resistance controlled, whereas absorption by raindrops is dispersed-phase controlled. Absorption by bubbles in the liquid phase is usually continuous-phase controlled.

7.2
Drag Force of a Bubble or Drop in a Creeping Flow (Hadamard's Flow)

7.2.1
Hadamard's Stream Function

In the previous section, we noted that small bubbles and drops are spherical in shape, and as flow around a fluid sphere is very slow, a creeping flow regime is established. Under these conditions, we can neglect the effect of the inertia term in comparison with that of the viscosity term in the equation of motion, as in the case of Stokes' flow (Section 6.1). Creeping flow around a fluid sphere is called *Hadamard's flow* in honor of J. S. Hadamard.

In Hadamard's flow, the equation of motion in terms of the stream function can be expressed as follows [7]:

Continuous phase:

$$\left[\frac{\partial^2}{\partial r^2} + \frac{\sin\theta}{r^2} \frac{\partial}{\partial \theta} \left(\frac{1}{\sin\theta} \frac{\partial}{\partial \theta} \right) \right]^2 \psi_c = 0 \tag{7.1}$$

Dispersed phase (inside bubbles or drops):

$$\left[\frac{\partial^2}{\partial r^2} + \frac{\sin\theta}{r^2} \frac{\partial}{\partial \theta} \left(\frac{1}{\sin\theta} \frac{\partial}{\partial \theta} \right) \right]^2 \psi_d = 0 \tag{7.2}$$

where ψ_c and ψ_d are the stream functions in the continuous phase and the dispersed phase, respectively.

The boundary conditions are as follows:
a) Uniform free stream velocity at large distances from a fluid sphere:

$$r = \infty : \psi_c/r^2 \to (U/2)\sin^2\theta \tag{7.3a}$$

b) No flow across the interface:

$$r = R : \psi_c = \psi_d \tag{7.3b}$$

c) Continuity of tangential velocity across the interface:

$$\frac{\partial \psi_c}{\partial r} = \frac{\partial \psi_d}{\partial r} \tag{7.3c}$$

d) Continuity of tangential stress across the interface:

$$\mu_c \frac{\partial}{\partial} \left(\frac{1}{r^2} \frac{\partial \psi_c}{\partial r} \right) = \mu_d \frac{\partial}{\partial} \left(\frac{1}{r^2} \frac{\partial \psi_d}{\partial r} \right) \tag{7.3d}$$

e) Continuity of normal stress across the interface:

$$p_c - 2\mu_c \frac{\partial}{\partial r} \left(\frac{1}{r^2 \sin\theta} \frac{\partial \psi_c}{\partial \theta} \right) + \frac{2\sigma}{R} = p_d - 2\mu_d \frac{\partial}{\partial r} \left(\frac{1}{r^2 \sin\theta} \frac{\partial \psi_d}{\partial \theta} \right) \tag{7.3e}$$

Solution of Eqs. (7.1) and (7.2) with the above boundary conditions, Eqs. (7.3a)–(7.3e), can easily be obtained by applying the method of separation of variables, details of which can be found in standard textbooks [7, 10].

The stream functions for the continuous phase and the dispersed phase are given as:

$$\psi_c = -\frac{U}{2} r^2 \sin^2\theta \left[1 - \frac{2 + 3(\mu_d/\mu_c)}{2\{1 + (\mu_d/\mu_c)\}} \left(\frac{R}{r} \right) + \frac{(\mu_d/\mu_c)}{2\{1 + (\mu_d/\mu_c)\}} \left(\frac{R}{r} \right)^3 \right] \tag{7.4}$$

$$\psi_d = \frac{1}{4\{1 + (\mu_d/\mu_c)\}} U r^2 \sin^2\theta \left\{ 1 - \left(\frac{R}{r} \right)^2 \right\} \tag{7.5}$$

where R is the radius of the fluid sphere [m], U is the velocity of the fluid far from the fluid sphere [m s^{-1}], and μ_c and μ_d are the viscosities of the continuous phase and the dispersed phase [Pa s], respectively.

Figure 7.2 shows the flow around a spherical particle ($\mu_d/\mu_c = \infty$), i.e. Stokes' flow, and that around a bubble ($\mu_d/\mu_c = 0$), i.e. Hadamard's flow, as calculated by means of Eqs. (7.4) and (7.5). The figures indicate there is a circulation flow inside a bubble, which is a characteristic of Hadamard's flow.

7.2.2
Drag Coefficients and Terminal Velocities of Small Drops and Bubbles [7, 10]

From the theoretical solution of the stream functions in Hadamard's flow, the drag coefficient of a small fluid sphere can be expressed as:

$$C_D = \frac{2 + 3(\mu_d/\mu_c)}{1 + (\mu_d/\mu_c)} \frac{8}{Re_p} \tag{7.6}$$

$$Re_p \equiv \rho_c D_p U / \mu_c \tag{7.7}$$

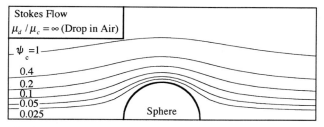

a) Flow around a Sphere

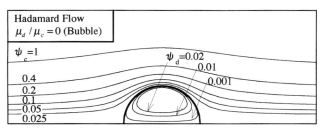

b) Flow around a Bubble

Fig. 7.2 Creeping flow around a spherical solid particle (Stokes' flow) and creeping flow around a bubble (Hadamard's flow).

From the force balance between the buoyancy force and the drag force under steady motion, we have the following equation.

$$\frac{\pi D_p^3}{6}(\rho_d - \rho_c)g = C_D \left(\frac{\pi D_p^2}{4}\right)\left(\frac{\rho_c U^2}{2}\right) \tag{7.8}$$

We can easily obtain the terminal velocity of a fluid sphere from Eqs. (7.6) and (7.8) as:

$$U = \frac{(\rho_d - \rho_c)g D_p^2}{6\mu_c}\left\{\frac{1+(\mu_d/\mu_c)}{2+3(\mu_d/\mu_c)}\right\} \tag{7.9}$$

The circulation velocity at the surface of a fluid sphere, $(v_\theta)_{r=R}$, is obtained from Eq. (7.4) as:

$$(v_\theta)_{r=R} = \left(\frac{1}{r\sin\theta}\frac{\partial \psi_c}{\partial r}\right)_{r=R} = \left\{\frac{1}{1+(\mu_d/\mu_c)}\right\}\frac{U\sin\theta}{2} \tag{7.10}$$

Equation (7.10) indicates that the circulation velocity becomes maximal at the equator of the fluid sphere and becomes zero at the two poles.

7.2 Drag Force of a Bubble or Drop in a Creeping Flow (Hadamard's Flow)

Motion of drops in the gas phase: For drops in the gas phase, the viscosity of the liquid is always very large in comparison with that of the gas phase, $\mu_d/\mu_c \gg 1$. Under these conditions, Eqs. (7.6), (7.9), and (7.10) reduce to the following forms:

$$C_D = C_{DST} = \frac{24}{Re_p} \tag{7.11}$$

$$U = U_{ST} = \frac{(\rho_d - \rho_c)gD_p^2}{18\mu_c} \tag{7.12}$$

$$(v_\theta)_{r=R} = 0 \tag{7.13}$$

where C_{DST} is the drag coefficient of a solid sphere in Stokes' flow as described in Chapter 6 and U_{ST} is the terminal velocity. This means that the drag coefficients and the terminal velocities coincide with those in Stokes' flow and that no circulation flow is expected at the surface of the drops. Therefore, the motion of small drops in the gas phase can be viewed as the motion of small spherical particles in Stokes' flow.

Motion of bubbles in the liquid phase: For bubbles in the liquid phase, the viscosity of the dispersed phase is negligibly small in comparison with that of the continuous phase, $\mu_d/\mu_c \ll 1$. Under these conditions, Eqs. (7.6), (7.9), and (7.10) reduce to the following forms:

$$U = \frac{(\rho_d - \rho_c)gD_p^2}{12\mu_c} = \frac{3}{2}U_{ST} \tag{7.14}$$

$$C_D = \frac{16}{Re_p} = \frac{2}{3}C_{DST} \tag{7.15}$$

The circulation velocity on the surface of the drops can be expressed as:

$$(v_\theta)_{r=R} = \left(\frac{U}{2}\right)\sin\theta \tag{7.16}$$

Motion of drops in the liquid phase: The terminal velocities, the drag coefficients, and the circulation velocities on the surfaces of drops in the liquid phase are seen to be intermediate between those of bubbles in the liquid phase and drops in the gas phase, and are dependent on the ratio of the viscosities of the dispersed and continuous phases μ_d/μ_c.

7.2.3
Motion of Small Bubbles in Liquids Containing Traces of Contaminants

If a liquid is contaminated with even a very small amount of a surface-active material, this is easily adsorbed at the interface between the gas and the liquid. The interface is immobilized and ultimately solidified by the adsorbed material. Under these circumstances, the surface circulation given by Eq. (7.10) will disappear and the motion of the surface-contaminated fluid spheres becomes similar to that of spherical particles in the fluid. Under ordinary laboratory conditions or in commercial plants, keeping the apparatus free from traces of contaminants is practically impossible, especially in cases in which water is used as the continuous phase. The fluid-fluid interface is very sensitive to small amounts of contaminants, and solidification of fluid spheres takes place. This means that the Hadamard flow regime described in this section is seldom encountered in actual problems.

7.3
Flow around an Evaporating Drop

7.3.1
Effect of Mass Injection or Suction on the Flow around a Spherical Particle

The evaporation of fuel sprays and the condensation of oil vapors on the surfaces of oil sprays are fundamental concerns in many industrial applications, such as the combustion of liquid fuels in internal combustion engines, the combustion of fuels in oil burners, and the recovery of hydrocarbon vapors from hydrocarbon polluted air with cold oil sprays. These phenomena are characterized by very high rates of mass transfer, and the flow around the drops is completely different from that under ordinary conditions. The drag coefficients and the heat and mass transfer of an evaporating or condensing drop are quite different from those of solid spheres. The high mass flux effect becomes operative.

As discussed in the previous section, the motion of a droplet in the gas phase can be regarded as the motion of a spherical particle. Thus, the flow around an evaporating or condensing drop in the gas phase can be approximated by the flow around a spherical particle with surface mass injection or suction.

P. Chuchottaworn, A. Fujinami, and K. Asano [3] presented a numerical approach to the effect of mass injection or suction on the drag coefficients of a spherical particle. Figure 7.3 shows the effect of uniform mass injection or suction on the flow around a spherical particle, where $\phi \, (\equiv v_s/U)$ is the mass injection (or suction) ratio, v_s is the normal component of the velocity at the surface of the particle [m s^{-1}], and U is the free stream velocity [m s^{-1}]. The figure indicates that the stream lines are pushed outward from the surface by mass injection ($\phi > 0$) and that the velocity gradient near the surface decreases along with the frictional drag. On the other hand, the stream lines are sucked to the surface by mass suction and the velocity gradient increases along with the frictional drag.

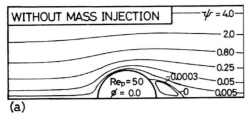
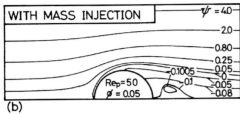
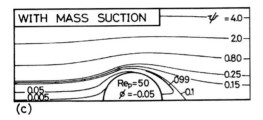

Fig. 7.3 Effect of uniform mass injection or suction on the flow around a spherical particle [3].

Figure 7.4 shows the effect of uniform mass injection or suction on the drag coefficients of a spherical particle. The numerical results are well correlated by the following equation:

$$\frac{C_D}{C_{D0}} = 1 - \frac{Re_p^{-0.043}}{1 + 0.328(\phi/C_{D0})^{-1}} \tag{7.17}$$

where C_{D0} is the drag coefficient of a spherical particle without mass injection or suction. P. Chuchottaworn and K. Asano [6] described a numerical approach to a more general case, where mass injection or suction is caused by evaporation of liquids or condensation of vapors, and they proposed the following correlation.

$$\frac{C_D}{C_{D0}} = \frac{1}{(1+B_M)^m} \tag{7.18}$$

$$m = 0.19 Sc^{-0.74}(1+B_M)^{-0.29} \tag{7.19}$$

where B_M is the transfer number for mass transfer as described in Section 4.7.

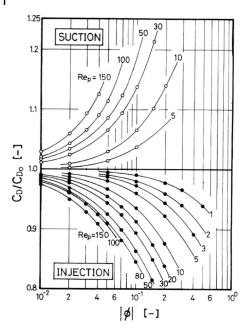

Fig. 7.4 Effect of uniform mass injection or suction on the drag coefficients of a spherical particle [3].

7.3.2
Effect of Mass Injection or Suction on Heat and Mass Transfer of a Spherical Particle

P. Chuchottaworn, A. Fujinami, and K. Asano [4] presented a numerical approach to the heat and mass transfer of an evaporating or condensing droplet in the gas phase and showed that the effect of mass injection or suction on the temperature and concentration distributions around a particle is similar to that on the velocity distribution shown in Fig. 7.3.

Figure 7.5 shows the effect of diffusive mass injection (evaporation) or suction (condensation) on the heat and mass transfer of an evaporating or condensing droplet in the gas phase. The ordinate, $g(B)$, represents the effect of high mass flux caused by evaporation or condensation, while the abscissa is the transfer number for heat and mass transfer, B_H and B_M, respectively. The bold solid line represents the correlation of numerical data for diffusive mass injection or suction.

For heat transfer:

$$g(B) \equiv \frac{Nu}{Nu_0} = \frac{1}{0.3 + 0.7(1+B)^{0.88}} \tag{7.20}$$

$$B = B_H \equiv \left(\frac{\overline{v_s}}{U}\right)\left(\frac{Re_p Pr}{Nu_0}\right) \tag{4.100}$$

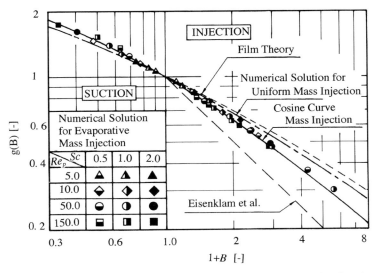

Fig. 7.5 Effect of mass injection or suction on the heat and mass transfer of a spherical particle.

For mass transfer:

$$g(B) \equiv \frac{Sh(1-\omega_s)}{Sh_0(1-\omega_s)} = \frac{1}{0.3 + 0.7(1+B)^{0.88}} \quad (7.21)$$

$$B = B_M \equiv \left(\frac{\overline{v_s}}{U}\right)\left(\frac{Re_p Sc}{Sh_0(1-\omega_s)}\right) = \frac{(\omega_s - \omega_\infty)}{(1-\omega_s)} \quad (4.101)$$

where Nu_0 and $Sh_0(1-\omega_s)$ are the Nusselt number and the dimensionless diffusional flux of a spherical particle, respectively.

For comparison, the figure also shows an approximate solution for uniform mass injection or suction by H. S. Mickley et al. [11]:

$$g(B) = \frac{\ln(1+B)}{B} \quad (7.22)$$

and an approximate solution by P. Eisenklam et al. [8]:

$$g(B) = \frac{1}{1+B} \quad (7.23)$$

Example 7.1
A hexane drop of diameter 1 mm is falling freely in air at 1 atm and 350 K at a velocity of 2.88 m s^{-1}. Calculate the rate of evaporation of the drop [kg s^{-1}] and the rate of decrease of the drop diameter [m s^{-1}] if the surface temperature of the drop is 280 K.

Solution

The physical properties of the system under the given conditions are as follows:

$\rho_{L280K} = 677$ kg m^{-3}, $\omega_s = 0.217$ [–],

$\rho_{Gs} = 1.46$ kg m^{-3}, $\mu_{Gs} = 1.68 \times 10^{-5}$ Pa s, $D_{Gs} = 7.55 \times 10^{-6}$ m^2 s^{-1},

$\rho_{G\infty} = 1.01$ kg m^{-3}, $\mu_{G\infty} = 2.06 \times 10^{-5}$ Pa s

There are big differences in the physical properties at the interface as compared to under free stream conditions. In such cases, previous experience suggests that the use of the Sherwood and Schmidt numbers for the physical properties at the interface and the Reynolds numbers for those under the free stream conditions is recommended for optimal correlation of the data.

$$Sc_s = \frac{(1.68 \times 10^{-5})}{(1.46)(7.55 \times 10^{-6})} = 1.53$$

$$Re_p = \frac{(1.01)(1.0 \times 10^{-3})(2.88)}{(2.06 \times 10^{-5})} = 141$$

The dimensionless diffusional flux in the absence of a high mass flux effect is given by use of the Ranz–Marshall correlation:

$$Sh_0 (1 - \omega_s) = 2 + (0.6)(1.53)^{1/3}(141)^{1/2} = 10.2$$

The high mass flux effect due to evaporation is estimated by means of Eq. (7.20):

$$B_M = (0.217)/(1 - 0.217) = 0.278$$

$$g(B_M) = \frac{1}{\{0.3 + 0.7(1 + 0.278)^{0.88}\}} = 0.856$$

This indicates that the high mass transfer rate of hexane reduces the mass transfer rate to 86% of that in the low flux model.

$$Sh(1 - \omega_s) = Sh_0(1 - \omega_s) g(B_M) = (10.2)(0.856) = 8.74$$

Mass flux at the surface of the evaporating drop:

$$N_A = \frac{(8.74)(1.46)(7.55 \times 10^{-6})(0.278)}{(1.0 \times 10^{-3})} = 2.66 \times 10^{-2} \text{ kg m}^{-2} \text{ s}^{-1}$$

Rate of evaporation of a single drop $= (2.66 \times 10^{-2})(3.14 \times 10^{-6})$
$= 8.35 \times 10^{-8}$ kg s^{-1}

From the material balance at the surface of the drop, the rate of decrease of the drop diameter is given by:

$$\frac{dD_p}{dt} = \left(\frac{-2N_A}{\rho_L}\right) = \frac{(2)(2.66 \times 10^{-2})}{(677)} = 7.85 \times 10^{-2} \text{ [mm/s]}$$

Note that this example is a rather approximate one, in which the surface temperature and the drop diameter are tentatively assumed. A more general solution to a problem in which the drop diameter and the surface temperature are changing simultaneously is shown in Example 7.2.

7.4
Evaporation of Fuel Sprays

7.4.1
Drag Coefficients, Heat and Mass Transfer of an Evaporating Drop

As discussed in the previous sections, a small drop in the gas phase behaves like a spherical particle. This means that drag coefficients and the heat and mass transfer of an evaporating drop in the gas phase can easily be obtained by applying the correlations for a spherical particle developed in Chapter 6, provided that the rate of evaporation is sufficiently low, as in the case of the evaporation of a water drop.

In the low to intermediate Reynolds number range of $Re_p < 1000$, the following correlations are applicable:

$$C_{D0} = \frac{24}{Re_p}(1 + 0.125\, Re_p^{0.72}) \tag{6.19}$$

$$Nu_{p0} = 2.0 + 0.6\, Pr^{1/3}\, Re_p^{1/2} \tag{6.33}$$

$$Sh_{p0}(1 - \omega_s) = 2.0 + 0.6\, Sc^{1/3}\, Re_p^{1/2} \tag{6.34}$$

For drops of volatile liquids such as gasoline, however, the rates of evaporation are very large and the flow around the surface of the drop is disturbed by the evaporation. Under these circumstances, we have to consider the effect of high mass flux, as discussed in the previous section.

The high mass flux effect on the drag coefficients:

$$\frac{C_D}{C_{D0}} = \frac{1}{(1 + B_M)^m} \tag{7.18}$$

$$m = 0.19\, Sc^{-0.74}(1 + B_M)^{-0.29} \tag{7.19}$$

The high mass flux effect on heat and mass transfer:

$$\frac{Nu}{Nu_0} = \frac{Sh(1-\omega_s)}{Sh_0(1-\omega_s)} = \frac{1}{0.3 + 0.7(1+B)^{0.88}} \quad (7.20), (7.21)$$

where $B = B_H$ for heat transfer and $B = B_M$ for mass transfer.

Figure 7.6 shows experimental results on the evaporation of a pentane drop in air at various ambient temperatures, as obtained by P. Chuchottaworn, A. Fujinami, and K. Asano [5]. The bold solid line in the figure represents the Ranz–Marshall correlation, Eq. (6.34), and the dashed lines are predictions by Eq. (7.21) based on the operating conditions. A systematic decrease in the observed data from the standard solid particle correlation by Ranz–Marshall may be attributed to the high mass flux effect caused by the evaporation of the pentane drop. The increase in transfer number for mass transfer, B_M, with increasing ambient temperature may be attributed to the increase in surface temperature and the increased saturated vapor pressure at higher ambient temperature. A good agreement between Eq. (7.21) and the observed data indicates that the correlations developed in Section 7.3 may be applied to the mass transfer of an evaporating drop.

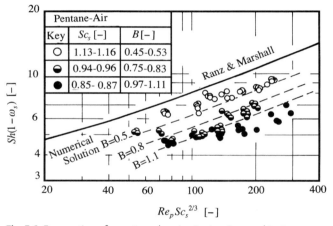

Fig. 7.6 Evaporation of a pentane drop in air at various ambient temperatures; comparison of the Ranz–Marshall correlation with experimental data obtained by P. Chuchottaworn et al. [5].

7.4.2
Behavior of an Evaporating Drop Falling Freely in the Gas Phase

The behavior of an evaporating volatile drop in the gas phase can easily be simulated by use of the equation of motion, and considering the mass and energy balance at the surface of the drop, where the flux correlations, Eqs. (6.19), (6.33), and (6.34), together with Eqs. (7.18) and (7.21), are used:

Equation of motion:

$$\frac{dv}{dt} = \left(1 - \frac{\rho_G}{\rho_L}\right)g - \frac{3}{4}\frac{C_D \rho_G}{D_p \rho_L}v^2 \qquad (7.24)$$

Mass balance on the surface of the drop:

$$\frac{dD_p}{dt} = \frac{-2N_A}{\rho_L} \qquad (7.25)$$

Heat balance on the surface of the drop:

$$\frac{dT_p}{dt} = \left(\frac{6}{\rho_L D_p c_L}\right)(q_G - \lambda_A N_A) \qquad (7.26)$$

K. Asano et al. [1] presented an experimental approach to the evaporation of a pentane drop falling freely in air. Figure 7.7 shows a comparison of the observed data on the drop diameter, the velocity of the drop, and the drop temperature with

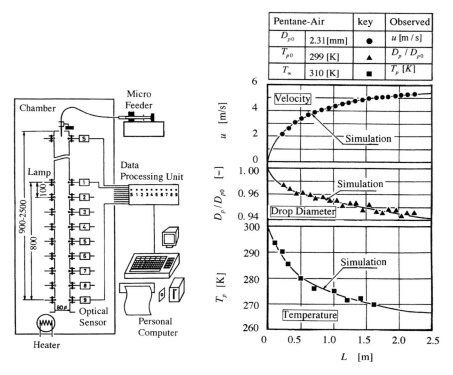

Fig. 7.7 Comparison of the observed data for an evaporating pentane drop falling freely in air with simulation taking into account the high mass flux effect [1].

the results of a simulation. Good agreement between the data and simulation is observed.

Example 7.2
A hexane drop of initial diameter 0.1 mm is falling freely in air at 1 atm and 323 K. Calculate the variation in the drop diameter, the drop temperature, and the velocity of the drop, if its initial velocity is 0 m s^{-1}.

Solution
Since the heat of vaporization is very large, the drop temperature will soon approach wet-bulb temperature. We assume that the physical properties of the system, except for the vapor pressure, are constant (at wet-bulb temperature). The following values are tentatively used in the calculation:

$\rho_{Air} = 1.093$ kg m^{-3}, $\mu_{Air} = 1.95 \times 10^{-5}$ Pa s

$D_{Hexane} = 6.70 \times 10^{-6}$ m^2 s^{-1}, $\rho_{Gs} = 1.47$ kg m^{-3}, $\kappa_{Gs} = 2.06 \times 10^{-2}$ W m^{-1} s^{-1}

$Pr = 0.81$, $Sc = 1.5$

$c_L = 2260$ J kg^{-1} K^{-1}, $\rho_L = 674$ kg m^{-3}, $\lambda_A = 3.80 \times 10^5$ J kg^{-1}

The vapor pressure of hexane is estimated by the use of Antoine's equation:

$$\log p = 8.99514 - \frac{1168.52}{T - 48.94} \tag{A}$$

The following finite difference forms of the governing equations are used in the calculation:

$$Z_{i+1} = Z_i + (v_i + v_{i+1}) \Delta t / 2 \tag{B}$$

$$v_{i+1} = v_i + \left(\frac{dv}{dt}\right)_i \Delta t \tag{C}$$

$$\left(\frac{dv}{dt}\right)_i = \left(1 - \frac{\rho_G}{\rho_L}\right) g - \frac{3}{4} \frac{C_D \rho_G}{D_p \rho_L} v_i^2 \tag{D}$$

$$C_{D,i} = (24/Re_p)(1 + 0.125 Re_p^{0.72}) \tag{E}$$

$$D_{p,j+1} = D_{p,j} + \left(\frac{dD_p}{dt}\right) \Delta t = D_{p,j} - \left(\frac{2 N_A}{\rho_L}\right) \Delta t \tag{F}$$

$$T_{p,i+1} = T_{p,i} + \left(\frac{dT_p}{dt}\right)_i \Delta t \tag{G}$$

The results of the calculation are shown in Tab. 7.1 and Fig. 7.8.

Table 7.1 Simulation of the evaporation of a hexane drop falling freely in air (Example 7.2)

t [s]	z [m]	D_p [mm]	v [m s^{-1}]	T_p [K]	C_D	Nu	Sh (1 – ω_s)	dv/dt	dT_p/dt
0.00	0.000	0.100	0.000	293.0	0.0	1.50	1.50	9.79	–875.2
0.01	0.000	0.097	0.094	284.1	48.6	1.98	2.06	4.48	–451.2
0.02	0.002	0.095	0.138	279.6	34.6	2.13	2.22	1.34	–153.6
0.03	0.003	0.093	0.152	278.0	32.4	2.17	2.27	0.08	–48.3
0.04	0.005	0.092	0.153	277.6	32.9	2.17	2.27	–0.35	–13.5
0.06	0.008	0.088	0.144	277.4	35.9	2.16	2.25	–0.52	0.3
0.08	0.010	0.085	0.134	277.4	40.1	2.13	2.22	–0.53	1.3
0.10	0.013	0.081	0.123	277.4	45.2	2.11	2.19	–0.53	1.4
0.12	0.015	0.077	0.112	277.5	51.6	2.08	2.16	–0.53	1.5
0.14	0.017	0.073	0.102	277.5	59.7	2.06	2.13	–0.53	1.5
0.16	0.019	0.069	0.091	277.5	70.3	2.03	2.10	–0.53	1.6
0.18	0.021	0.064	0.081	277.6	84.3	2.01	2.07	–0.53	1.7
0.20	0.023	0.060	0.070	277.6	104.0	1.98	2.03	–0.53	1.8
0.22	0.024	0.055	0.059	277.6	132.6	1.95	2.00	–0.53	1.9
0.24	0.025	0.050	0.049	277.7	177.3	1.92	1.96	–0.53	2.1
0.26	0.026	0.044	0.038	277.7	254.1	1.89	1.92	–0.52	2.3
0.28	0.026	0.037	0.028	277.8	406.5	1.86	1.88	–0.51	2.5
0.30	0.027	0.030	0.018	277.8	796.8	1.82	1.84	–0.49	2.9
0.32	0.027	0.020	0.008	277.9	2578.0	1.79	1.80	–0.44	3.5
0.33	0.027	0.014	0.004	277.9	8349.6	1.76	1.77	–0.34	3.8
0.34	0.027	0.004	0.000	277.9					

The figure highlights some interesting facts about an evaporating drop. The drop temperature decreases rapidly in the early stages of the motion due to the effect of the high latent heat of vaporization of the liquid, but soon approaches a constant value (wet-bulb temperature). The drop diameter decreases with time. The drop velocity soon attains its maximum value and then decreases gradually as the drop diameter decreases. A characteristic of the motion of an evaporating drop is that it does not have a terminal velocity, in contrast to the situation with a solid particle.

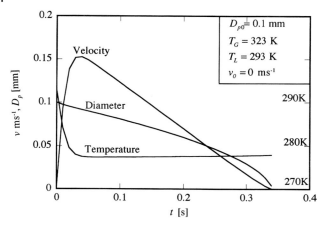

Fig. 7.8 Variations in the velocity, diameter, and temperature of an evaporating hexane drop falling freely in air.

7.5
Absorption of Gases by Liquid Sprays

In the absorption of sparingly soluble gases by liquid sprays, such as the absorption of carbon dioxide by water sprays, we can neglect the continuous-phase mass transfer resistances compared to the dispersed-phase mass transfer resistances and the phenomenon can be approximated as dispersed-phase controlled. Moreover, as described in Section 7.2, the motion of drops in the gas phase, where $\mu_d \gg \mu_c$, can be approximated by the motion of solid spheres in the gas phase and the *solid-sphere penetration model* is applicable to mass transfer in this case.

The diffusion equation for the dispersed phase can be expressed as follows:

$$\frac{\partial \omega}{\partial t} = \frac{D_d}{r^2} \frac{\partial}{\partial r}\left(r^2 \frac{\partial \omega}{\partial r}\right) \tag{7.27}$$

The initial and the boundary conditions are:

$$t = 0: \quad \omega = \omega_0 \tag{7.28a}$$

$$r = R: \quad \omega = \omega_s \tag{7.28b}$$

where D_d [m^2 s^{-1}] is the diffusivity of the solute gas in the liquid, R [m] is the radius of the drop, and ω_0 [–] and ω_s [–] are the initial concentration of the solute gas in the drop and the equilibrium concentration at the surface of the drop, respectively.

The analytical solution of Eq. (7.27), the details of which are presented in a standard textbook [2], is given by the following equation:

$$\theta_c \equiv \frac{\omega - \omega_0}{\omega_s - \omega_0} = 1 + \frac{2}{\pi}\left(\frac{R}{r}\right)\sum_{n=1}^{\infty}\frac{(-1)^n}{n}\sin\left(\frac{n\pi r}{R}\right)\exp\left(\frac{-D_d \pi^2 t}{R^2}n^2\right) \quad (7.29)$$

The average concentration of the solute gas in the drop is obtained by integration of Eq. (7.29) as:

$$\overline{\theta_c} \equiv \int_0^R 4\pi r^2 \theta_c dr/(4\pi R^3/3) = 1 - \frac{6}{\pi^2}\sum_{n=1}^{\infty}\frac{1}{n^2}\exp(-4n^2\pi^2 Fo) \quad (7.30)$$

$$Fo \equiv D_d t/D_p^2 \quad (7.31)$$

where Fo is the Fourier number for mass transfer.

The average rate of mass transfer over a time interval of t_c [s], $\overline{N_A}$ [kg m^{-2} s^{-1})], is given by the following equation:

$$\overline{N_A} = \left(\frac{1}{4\pi R^2 t_c}\right)\left\{\frac{4\pi R^3 \rho_d(\overline{\omega} - \omega_0)}{3}\right\}$$

$$= \left(\frac{\rho_d D_p}{6 t_c}\right)(\omega_s - \omega_0)\left\{1 - \frac{6}{\pi^2}\sum_{n=1}^{\infty}\frac{1}{n^2}\exp(-4n^2\pi^2 Fo)\right\} \quad (7.32)$$

Rearrangement of Eq. (7.32) into dimensionless form gives the dispersed-phase Sherwood numbers as:

$$Sh_d \equiv \frac{\overline{N_A} D_p}{\rho_d D_d(\omega_s - \omega_0)} = \frac{1}{6 Fo}\left\{1 - \frac{6}{\pi^2}\sum_{n=1}^{\infty}\frac{1}{n^2}\exp(-4n^2\pi^2 Fo)\right\} \quad (7.33)$$

Figure 7.9 shows a comparison of the theoretical values obtained by means of Eq. (7.33) with the observed data for absorption of carbon dioxide from air/carbon dioxide mixtures of various concentrations with water sprays as obtained by I. Taniguchi et al. [12]. The solid line in the figure represents Eq. (7.33) and good agreement between the calculated and experimental data is observed.

Example 7.3
A water drop of diameter 1 mm is falling freely in a gas mixture of 20% CO_2 and 80% air at 1 atm and 294 K. Calculate the amount of carbon dioxide absorbed by this single drop at a distance of 10 m from the starting point, if its initial velocity is 0 m s^{-1} and the absolute humidity of the gas is 0.01.

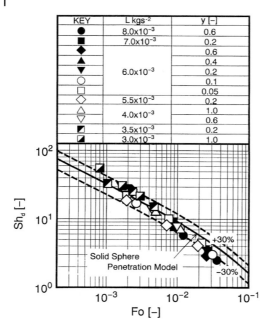

Fig. 7.9 Absorption of carbon dioxide by water sprays; comparison of the solid-sphere penetration model with experimental data obtained by Taniguchi et al. [12].

Solution

We assume that the temperature of the drop will soon approach the wet-bulb temperature of the gas.

From the humidity chart, the wet-bulb temperature of the gas is estimated to be 290 K. The physical properties of the system under the given operating conditions are estimated as:

$D_L = 1.50 \times 10^{-9} \text{ m}^2 \text{ s}^{-1}$

$\rho_L = 1000 \text{ kg m}^{-3}$

$H = 131.4 \text{ MPa (Henry's constant)}$

$x_s = (0.2)(101.325)/(1.314 \times 10^5) = 1.54 \times 10^{-4}$

$\omega_s = (1.54 \times 10^{-4})(44)/\{(1 - 1.54 \times 10^{-4})(18) + (1.54 \times 10^{-4})(44)\}$
$= 3.76 \times 10^{-4}$

Since the vapor pressure of water at the surface of the drop is sufficiently low and the rate of evaporation is negligibly small, we will assume that the drop diameter is essentially constant over an interval of 10 m. Thus, the contact time of the drop can be estimated by numerical integration of Eq. (7.24). Table 7.2 shows the results of a simulation of the motion of a water drop falling freely in the gas phase.

7.5 Absorption of Gases by Liquid Sprays

Table 7.2 Simulation of the motion of a water drop falling freely in an air/carbon dioxide mixture; time, velocity, and distance travelled by the drop.

t [s]	v [m s^{-1}]	y [m]	Re$_p$	C$_D$
0.00	0.00	0.00	0.0	0.000
0.20	1.96	0.20	129.9	0.953
0.40	3.26	0.72	216.2	0.777
0.60	3.73	1.42	247.5	0.738
0.80	3.84	2.17	254.7	0.730
1.00	3.86	2.95	256.0	0.729
1.20	3.87	3.72	256.2	0.729
1.40	3.87	4.49	256.3	0.729
1.60	3.87	5.26	256.3	0.728
1.80	3.87	6.04	256.3	0.728
2.00	3.87	6.81	256.3	0.728
2.20	3.87	7.58	256.3	0.728
2.40	3.87	8.36	256.3	0.728
2.60	3.87	9.13	256.3	0.728
2.80	3.87	9.90	256.3	0.728
3.00	3.87	10.68	256.3	0.728

Interpolation of the last two rows of Tab. 7.2 gives the exposure time of the water drop as:

$t_c = 2.83$ s

$$Fo = \frac{(1.50 \times 10^{-9})(2.83)}{(1.0 \times 10^{-3})^2} = 4.25 \times 10^{-3}$$

Substitution of the above value into Eq. (7.33) gives:

$Sh_d = 15.25$

The average rate of absorption of the solute gas is given by:

$$\overline{N_A} = \frac{(15.25)(1000)(1.50 \times 10^{-9})(3.76 \times 10^{-4})}{(1.0 \times 10^{-3})} = 8.60 \times 10^{-6} \text{ kg m}^{-2} \text{ s}^{-1}$$

The average concentration in the drop at a distance of 10 m is given by:

$$\omega_{av} = \frac{6\,\overline{N_A}\,t_c}{\rho_d D_p} = \frac{(6)(8.60 \times 10^{-6})(2.85)}{(1000)(1.0 \times 10^{-3})} = 1.47 \times 10^{-4}$$

$\omega_{av}/\omega_s = (1.47 \times 10^{-4})/(3.76 \times 10^{-4}) = 0.39$

The calculation shows that the average concentration of the solute gas becomes about 40% of the equilibrium concentration.

7.6
Mass Transfer of Small Bubbles or Droplets in Liquids

7.6.1
Continuous-Phase Mass Transfer of Bubbles and Droplets in Hadamard Flow

The rates of mass transfer of bubbles and drops in liquids during the formation period are very large in comparison with those during steady motion. However, the behavior of bubbles or drops is quite complicated and no quantitative results have been obtained. For this reason, we will restrict our discussions to mass transfer during a steady motion. Schmidt numbers in the liquid phase are usually very large, ranging from several hundreds to twenty to thirty thousands. As described in Section 3.5, the concentration boundary layer lies far inside the velocity boundary layer, or very near to the surface in the case of bubbles or droplets in the liquid phase.

V. G. Levich [10] presented a theoretical approach to this problem by approximating the diffusion equation in Hadamard flow by the following equation:

$$\frac{v_\theta}{R}\frac{\partial \omega}{\partial \theta} = D\frac{\partial^2 \omega}{\partial y^2} \tag{7.34}$$

Approximating the stream functions for the continuous phase by the following equation:

$$v_\theta = \frac{-1}{r\sin\theta}\frac{\partial \psi}{\partial r} \approx \frac{-3}{2}\left(\frac{r}{R}-1\right)U\sin\theta \tag{7.35}$$

the following equation is obtained by considering the boundary conditions:

$$Sh_c(1-\omega_s) = 0.651\left(\frac{UD_p}{D_c}\right)^{1/2}\left\{\frac{1}{1+(\mu_d/\mu_c)}\right\} \tag{7.36}$$

For small bubbles rising in the liquid ($\mu_c \gg \mu_d$), Eq. (7.36) reduces to the following equation:

$$Sh_c(1-\omega_s) = 0.651\left(\frac{UD_p}{D_c}\right)^{1/2} \tag{7.37}$$

Equation (7.37) indicates that continuous-phase mass transfer for bubbles in Hadamard flow shows different trends from that in the case of solid particles, Eq. (6.39), due to the effect of the circulation flow inside the bubbles. If, however, the systems are contaminated with small amounts of surface-active materials, the surfaces of the bubbles or drops are immobilized by the adsorbed material. As a result, the mass transfer of bubbles and drops in such a contaminated liquid will behave like that of solid particles, Eq. (6.39).

7.6.2
Dispersed-Phase Mass Transfer of Drops in Hadamard Flow

R. Kronig and J. C. Brink [9] derived the following approximate solution for the dispersed-phase mass transfer of drops in Hadamard flow:

$$Sh_d = \frac{1}{6F_0}\{1 - 0.6534\exp(-107.4F_0) - 0.2\exp(-629F_0)\} \tag{7.38}$$

For the reason outlined above, Eq. (7.38) is not applicable for the mass transfer of drops in a liquid contaminated with a surface-active agent. In such cases, the solid-sphere penetration model, Eq. (7.33), is applicable instead.

Figure 7.10 shows a comparison of the two dispersed-phase mass transfer models, the Kronig–Brink model and the solid-sphere penetration model.

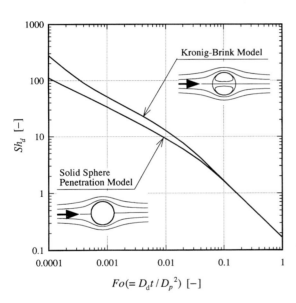

Fig. 7.10 Dispersed-phase mass transfer of drops in liquids; comparison of the solid-sphere penetration model with the Kronig–Brink model.

7.6.3
Mass Transfer of Bubbles or Drops of Intermediate Size in the Liquid Phase

As described in Section 7.1, bubbles and drops of intermediate size tend to be deformed from a spherical shape to oblate spheroidal and even to mushroom-like shapes. The motion of bubbles or drops of intermediate size can be quite unstable, and sometimes even oscillatory. The transport phenomena of intermediate sized bubbles and drops are thus very complicated, and no theoretical description has ever been attempted.

Example 7.4

Air bubbles are introduced into water at 20 °C from the bottom of a container of depth 40 cm. Calculate the rate of absorption of oxygen by a single bubble, assuming that the diameter of the air bubble is 1 mm and that the average rising velocity of the bubble is 30% of the theoretical terminal velocity in Hadamard flow.

Solution

If we assume that the surface temperature of the bubble is the same as that of the water (20 °C), the physical properties of the system are given by:

$$\rho_c = 1000 \text{ kg m}^{-3}, \mu_c = 1.0 \text{ mPa s}, D_c = 2.0 \times 10^{-9} \text{ m}^2 \text{ s}^{-1}$$
$$\rho_d = 1.20 \text{ kg m}^{-3}, H = 4052 \text{ MPa}$$

The equilibrium oxygen concentration on the surface of the bubble at the bottom of the container:

$$P_{Bottom} = 101325 + (1000)(9.81)(0.4) = 105249 \text{ Pa}$$
$$x_s = (1.05249 \times 10^5)(0.21)/(4.052 \times 10^9) = 5.45 \times 10^{-6}$$
$$\omega_s = (5.45 \times 10^{-6})(32)/(1.0)(18) = 9.69 \times 10^{-6}$$

The average rising velocity of the bubble is calculated by means of Eq. (7.14) as:

$$U_{Hadamard} = \frac{(1000 - 1.2)(9.81)(1.0 \times 10^{-3})^2}{(12)(1.0 \times 10^{-3})} = 0.817 \text{ m s}^{-1}$$
$$U = (0.817)(0.3) = 0.245 \text{ m s}^{-1}$$

The continuous-phase mass transfer of an air bubble is given by Eq. (7.37):

$$Sh_c (1 - \omega_s) = (0.651)\left(\frac{(0.245)(1.0 \times 10^{-3})}{(2.0 \times 10^{-9})}\right)^{1/2} = 228$$

$$N_A = \frac{(228)(1000)(2.0 \times 10^{-9})(9.68 \times 10^{-9})}{(1.0 \times 10^{-3})}$$
$$= 4.41 \times 10^{-6} \text{ kg m}^{-2} \text{ s}^{-1}$$

The surface area of a single bubble:

$$S = (3.14)(1.0 \times 10^{-3})^2 = 3.14 \times 10^{-6} \text{ m}^2$$

Contact time of the bubble:

$$t_c = (0.4)/(0.245) = 1.63 \text{ s}$$

The amount of oxygen absorbed by a single bubble is given by:

$$Q = (4.41 \times 10^{-6})(3.14 \times 10^{-6})(1.63)$$
$$= 2.26 \times 10^{-11} \text{ kg bubble}^{-1}$$

References

1 K. Asano, I. Taniguchi, K. Maeda, H. Kosuge, "Simultaneous Measurements of Drag Coefficients and Mass Transfer of a Volatile Drop Falling Freely", *Journal of Chemical Engineering of Japan*, **21**, [4], 387–393 (1988).

2 H. S. Carslaw and J. C. Jeager, "*Conduction of Heat in Solid*", 2nd Edition, p. 233–234, Oxford University Press (1959).

3 P. Chuchottaworn, A. Fujinami, and K. Asano, "Numerical Analysis of the Effect of Mass Injection and Suction on Drag Coefficients of a Sphere", *Journal of Chemical Engineering of Japan*, **16**, [1], 18–24 (1983).

4 P. Chuchottaworn, A. Fujinami, and K. Asano, "Numerical Analysis of Heat and Mass Transfer from a Sphere with Surface Mass Injection and Suction", *Journal of Chemical Engineering of Japan*, **17**, [1], 1–7 (1984).

5 P. Chuchottaworn, A. Fujinami, and K. Asano, "Experimental Study of Evaporation of a Volatile Pendant Drop under High Mass Flux Conditions", *Journal of Chemical Engineering of Japan*, **17**, [1], 7–13 (1984).

6 P. Chuchottaworn and K. Asano, "Calculation of Drag Coefficients of an Evaporating or a Condensing Droplet", *Journal of Chemical Engineering of Japan*, **18**, [1], 91–94 (1985).

7 R. Clift, J. R. Grace, and M. E. Weber, "*Bubbles, Drops, and Particles*", p. 30–68, Academic Press (1978).

8 P. Eisenklam, S. A. Arunachalam, and J. A. Weston, "Evaporation Rates and Drag Coefficients of Burning Drops", *Eleventh Symposium (International) on Combustion*, pp. 715–727, Combustion Institute, Pittsburg (1967).

9 R. Kronig and J. C. Brink, "On the Theory of Extraction from Falling Droplets", *Applied Science Research*, **A2**, 142–154 (1950).

10 V. G. Levich, "*Physicochemical Hydrodynamics*", p. 80–87, Prentice-Hall (1960).

11 H. S. Mickley, R. C. Ross, A. L. Squyers, and W. E. Stewart, *NACA Technical Note*, **3208** (1954).

12 I. Taniguchi, Y. Takamura, and K. Asano, "Experimental Study of Gas Absorption with Spray Column", *Journal of Chemical Engineering of Japan*, **30**, [3], 427–433 (1997).

8
Turbulent Transport Phenomena

8.1
Fundamentals of Turbulent Flow

8.1.1
Turbulent Flow

As discussed in Chapter 3, there are two fundamental types of fluid flow, namely laminar and turbulent. The transition from laminar to turbulent flow takes place when the Reynolds number exceeds a certain critical value. The critical Reynolds number for a flow inside a circular pipe is in the range 2000 to 2300, while for a flow in a boundary layer along a flat plate it is in the range 3.5×10^5 to 1×10^6.

Figure 8.1 shows a schematic representation of the fluctuations in the local velocity at a given point in a turbulent flow, as monitored by a highly sensitive sensor such as a hot-wire anemometer. In the figure, only the x-component of the velocity is shown for the sake of simplicity, but similar trends are observed for the y- and z-components. The irregular fluctuation of the velocity in all directions is the essential characteristic of turbulent flow.

The components of the local velocity at a given point (x, y, z) in a turbulent flow can be expressed as the sums of the time-averaged velocities ($\bar{u}, \bar{v}, \bar{w}$) and the fluctuation velocities (u', v', w') by way of the following equations:

$$u = \bar{u} + u' \tag{8.1a}$$

$$v = \bar{v} + v' \tag{8.1b}$$

$$w = \bar{w} + w' \tag{8.1c}$$

The pressures, temperatures, and concentrations in the turbulent flow also show irregular fluctuations, and are likewise expressed as the sums of the time-averaged and fluctuation terms.

$$p = \bar{p} + p' \tag{8.2a}$$

$$T = \bar{T} + T' \tag{8.2b}$$

$$\omega = \bar{\omega} + \omega' \tag{8.2c}$$

Mass Transfer. From Fundamentals to Modern Industrial Applications. Koichi Asano
Copyright © 2006 WILEY-VCH Verlag GmbH & Co. KGaA, Weinheim
ISBN: 3-527-31460-1

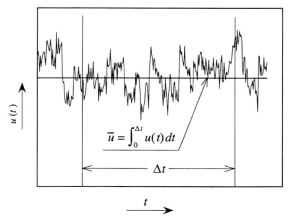

Figure 8.1 Velocity fluctuations in turbulent flow

8.1.2
Reynolds Stress

Although the governing equations described in Chapter 3 may also be applicable to transport phenomena in a turbulent flow, obtaining exact solutions of these equations is practically impossible due to the irregular fluctuations of the velocities, temperatures, and concentrations in the turbulent flow. For this reason, discussions on turbulent transport phenomena are usually based on time-averaged velocities, temperatures, and concentrations.

Let us consider the continuity equation in terms of the time-averaged velocity. Taking into account the fact that the components of the time-averaged velocity fluctuations ($\overline{u'}$, $\overline{v'}$, $\overline{w'}$) are always zero:

$$\overline{u'} = \overline{v'} = \overline{w'} = 0 \tag{8.3}$$

and substituting Eqs. (8.1a)–(8.1c) into Eq. (3.3), we have the following equation:

$$\frac{\partial \overline{u}}{\partial x} + \frac{\partial \overline{v}}{\partial y} + \frac{\partial \overline{w}}{\partial z} = 0 \tag{8.4}$$

This means that the time-averaged continuity equation has the same form as that for a laminar flow.

In the case of the equation of motion, however, the situation is quite different due to the effect of the quadratic velocity terms. Consider, for example, the x-component of the time-averaged equation of motion by substituting Eqs. (8.1) into Eq. (3.15a).

$$\bar{u}\frac{\partial \bar{u}}{\partial x} + \bar{v}\frac{\partial \bar{u}}{\partial x} + \bar{w}\frac{\partial \bar{u}}{\partial x} = g_x - \frac{\partial \bar{P}}{\partial x} + \left(\frac{\mu}{\rho}\right)\left(\frac{\partial^2 \bar{u}}{\partial x^2} + \frac{\partial^2 \bar{u}}{\partial y^2} + \frac{\partial^2 \bar{u}}{\partial z^2}\right)$$

$$+ \left(\frac{1}{\rho}\right)\left(\frac{\partial \tau_{turb,xx}}{\partial x} + \frac{\partial \tau_{turb,yx}}{\partial y} + \frac{\partial \tau_{turb,zx}}{\partial z}\right) \quad (8.5)$$

The fourth term on the right-hand side of Eq. (8.5) is due to the time-average of the quadratic terms of the velocity fluctuations, and does not become zero even after a time-averaging operation. Similar expressions are obtained for the y- and z-components. An important fact about the time-averaged equation of motion is that it is different from that for a laminar flow in that it contains an extra term for the apparent stress due to quadratic velocity fluctuations, $\tau_{turb,ij}$. This apparent stress due to quadratic terms of the velocity fluctuation, $\tau_{turb,ij}$, is called the Reynolds stress.

$$\begin{pmatrix} \tau_{turb,xx} & \tau_{turb,yx} & \tau_{turb,zx} \\ \tau_{turb,xy} & \tau_{turb,yy} & \tau_{turb,zy} \\ \tau_{turb,xz} & \tau_{turb,yz} & \tau_{turb,zz} \end{pmatrix} = -\rho \begin{pmatrix} \overline{u'^2} & \overline{u'v'} & \overline{u'w'} \\ \overline{v'u'} & \overline{v'^2} & \overline{v'w'} \\ \overline{w'u'} & \overline{w'v'} & \overline{w'^2} \end{pmatrix} \quad (8.6)$$

8.1.3
Eddy Heat Flux and Diffusional Flux

The time-averaged energy and diffusion equations can also be obtained from Eqs. (3.8) and (3.14) and are expressed as follows:
The time-averaged energy equation:

$$\bar{u}\frac{\partial \bar{T}}{\partial x} + \bar{v}\frac{\partial \bar{T}}{\partial y} + \bar{w}\frac{\partial \bar{T}}{\partial y} = \left(\frac{\kappa}{\rho c_p}\right)\left(\frac{\partial^2 \bar{T}}{\partial x^2} + \frac{\partial^2 \bar{T}}{\partial y^2} + \frac{\partial^2 \bar{T}}{\partial y^2}\right) +$$

$$+ \left(\frac{1}{\rho c_p}\right)\left(\frac{\partial q_{turb,x}}{\partial x} + \frac{\partial q_{turb,y}}{\partial y} + \frac{\partial q_{turb,z}}{\partial z}\right) \quad (8.7)$$

where $q_{turb,x}$, $q_{turb,y}$, $q_{turb,z}$ are the turbulent heat fluxes (or eddy heat fluxes):

$$\begin{pmatrix} q_{turb,x} \\ q_{turb,y} \\ q_{turb,z} \end{pmatrix} = -\rho c_p \begin{pmatrix} \overline{u'T'} \\ \overline{v'T'} \\ \overline{w'T'} \end{pmatrix} \quad (8.8)$$

The time-averaged diffusion equation:

$$\bar{u}\frac{\partial \bar{\omega}}{\partial x} + \bar{v}\frac{\partial \bar{\omega}}{\partial y} + \bar{w}\frac{\partial \bar{\omega}}{\partial z} = D\left(\frac{\partial^2 \bar{\omega}}{\partial x^2} + \frac{\partial^2 \bar{\omega}}{\partial y^2} + \frac{\partial^2 \bar{\omega}}{\partial z^2}\right) +$$

$$+ \left(\frac{1}{\rho}\right)\left(\frac{\partial J_{turb,x}}{\partial x} + \frac{\partial J_{turb,y}}{\partial y} + \frac{\partial J_{turb,z}}{\partial z}\right) \quad (8.9)$$

where $J_{turb,x}$, $J_{turb,y}$, $J_{turb,z}$ are the turbulent diffusional fluxes (or eddy diffusional fluxes):

$$\begin{pmatrix} J_{turb,x} \\ J_{turb,y} \\ J_{turb,z} \end{pmatrix} = -\rho \begin{pmatrix} \overline{u'\omega'} \\ \overline{v'\omega'} \\ \overline{w'\omega'} \end{pmatrix} \tag{8.10}$$

8.1.4
Eddy Transport Properties

In the following, we will discuss two-dimensional flow for the sake of simplicity. Shear stress in a turbulent flow is composed of two parts, namely the shear stress due to the molecular transport mechanism and the apparent stress due to quadratic velocity fluctuation terms or eddy terms:

$$\tau = -\rho(v+\varepsilon)\frac{\partial \overline{u}}{\partial y} \tag{8.11}$$

$$\tau_{turb} = -\rho\overline{u'v'} \equiv -\varepsilon\frac{\partial \overline{u}}{\partial y} \tag{8.12}$$

where τ_{turb} represents the Reynolds stress, $v(\equiv \mu/\rho)$ is the kinematic viscosity [m² s⁻¹], and ε is the eddy kinematic viscosity [m² s⁻¹].

In a similar way, the turbulent heat flux (or eddy heat flux) and the turbulent diffusional flux (or eddy diffusional flux) can be expressed by the following equations:

Heat flux:

$$q = -\rho c_p(\alpha + \varepsilon_H)\frac{\partial \overline{T}}{\partial y} \tag{8.13}$$

$$q_{turb} = -\rho c_p \overline{T'v'} \equiv -\rho c_p \varepsilon_H \frac{\partial \overline{T}}{\partial y} \tag{8.14}$$

Diffusional flux:

$$J = -\rho(D + \varepsilon_D)\frac{\partial \overline{\omega}}{\partial y} \tag{8.15}$$

$$J_{turb} = -\rho\overline{\omega'v'} \equiv -\rho\varepsilon_D \frac{\partial \overline{\omega}}{\partial y} \tag{8.16}$$

where D is the diffusivity [m² s⁻¹], J_{turb} is the turbulent diffusional flux [kg m⁻² s⁻¹], q_{turb} is the turbulent heat flux [W m⁻²], $\alpha(\equiv \kappa/\rho c_p)$ is the thermal diffusivity [m² s⁻¹],

ε_D is the turbulent diffusivity (eddy diffusivity) [m² s⁻¹], and ε_H is the turbulent thermal diffusivity (eddy thermal diffusivity) [m² s⁻¹].

Although the apparent relationships between the turbulent fluxes and the driving forces are similar to the molecular relationships, an important fact about the turbulent transport properties is that they are not true physical properties dependent only on the molecular transport mechanisms. On the other hand, they are dependent on the distance from the wall as well as on the flow conditions.

8.1.5
Mixing Length Model

L. Prandtl [10] proposed the important concept of mixing length in turbulent transport phenomena, in analogy to the concept of mean free path in the kinetic theory of gases.

Let us consider a two-dimensional flow and a change of momentum of a small element of fluid at the plane **B**, as shown in Fig. 8.2. We assume that small eddies at planes **A** and **C**, which are l [m] from plane **B** at time t, will reach plane **B** after δt [s] by the motion of the transverse velocity fluctuation v' [m s⁻¹] without a loss of momentum in the x-direction.

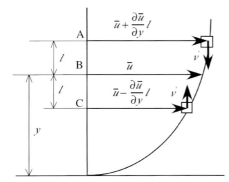

Figure 8.2 Mixing length model

The difference in the velocity at plane **A** is given by:

$$\Delta u_1 = l \left(\frac{\partial \bar{u}}{\partial y} \right)$$

The difference in the velocity at plane **C** is given by:

$$\Delta u_2 = -l \left(\frac{\partial \bar{u}}{\partial y} \right)$$

The net change in momentum in the x-direction at plane **B** due to the eddy transport mechanism is then given by the following equation:

$$\{|\Delta u_1| + |\Delta u_2|\} = l\left(\frac{\partial \bar{u}}{\partial y}\right) \tag{8.17}$$

The *x*-component of the velocity fluctuation is related to the *y*-component thereof by the continuity equation, and we may assume the following equation:

$$|u'| \approx |v'| = l\left(\frac{\partial \bar{u}}{\partial y}\right) \tag{8.18}$$

Substituting Eq. (8.18) into Eq. (8.12), we have the following equation:

$$|\tau_{turb}| \equiv \rho \overline{u'v'} = \rho l^2 \left(\frac{\partial \bar{u}}{\partial y}\right)^2 \tag{8.19}$$

where l is referred to as Prandtl's *mixing length*.

J. Nikuradse proposed the following correlation for the mixing length in a turbulent flow inside a smooth circular pipe for the Reynolds number range $Re > 10^5$ on the basis of his precise measurements on the velocity distribution [12].

$$\left(\frac{l}{R}\right) = 0.14 - 0.08\left(1 - \frac{y}{R}\right)^2 - 0.06\left(1 - \frac{y}{R}\right)^4 \tag{8.20}$$

Equation (8.20) can be approximated by the following equation for turbulent flow near the wall region ($y/R \leq 0.05$):

$$l = 0.4\, y \tag{8.21}$$

E. R. van Driest [3] proposed a modified form of Eq. (8.21) by introducing a *dumping factor* concept based on the following equation:

$$l = 0.4\, y(1 - \exp(-y^+/26)) \tag{8.22}$$

where $y^+ (\equiv \rho v^* y/\mu)$ is the dimensionless distance from the wall, the details of which are discussed in Section 8.2.

8.2
Velocity Distribution in a Turbulent Flow inside a Circular Pipe and Friction Factors

8.2.1
1/*n*-th Power Law

For velocity distributions in a turbulent flow inside a circular pipe, the following equation is empirically known:

$$\left(\frac{\bar{u}}{U_{max}}\right) = \left(\frac{y}{R}\right)^{1/n} \tag{8.23}$$

This equation is known as the *1/n-th power law*, where U_{max} is the maximum velocity at the center of the pipe [m s^{-1}], R is the radius of the pipe [m], and y is the distance from the wall [m]. According to the detailed and accurate measurements by J. Nikuradse [9], the numerical value of the exponent n in the above equation lies in the range from 6 to 10. For the practically important Reynolds number range of $2 \times 10^4 < Re < 10^5$, the numerical value of the exponent becomes $n = 7$, and the equation is called the *1/7-th power law*.

Although the *1/n-th power law* can adequately represent the velocity distribution in almost all regions of a circular pipe, it is not valid at the two ends. At the wall ($y = 0$), the velocity gradient becomes infinite, while at the center of the pipe ($y = R$), the velocity gradient is finite and not zero.

The average velocity over the cross-section of the pipe can be easily calculated by integrating Eq. (8.23) with respect to y from $y = 0$ to $y = R$:

$$\frac{U_m}{U_{max}} = \frac{1}{\pi R^2} \int_0^R 2\pi r \left(1 - \frac{r}{R}\right)^{1/n} dr = \frac{2n^2}{(n+1)(2n+1)}$$

8.2.2
Universal Velocity Distribution Law for Turbulent Flow inside a Circular Pipe

According to measurements by H. Reichardt [11], the Reynolds stress increases rapidly and becomes maximal in the region very near to the wall then decreases gradually and linearly as the distance from the wall increases. If we confine our discussions to the flow in the region near to the wall, the turbulent shear stress can be approximated by the following equation:

$$\tau_{turb} = \rho \overline{u'v'} \approx \rho \overline{v'^2} \approx \tau_w \tag{8.25}$$

Rearranging Eq. (8.25) by use of the definition of the friction factor:

$$f \equiv \frac{\tau_w}{(\rho U_m^2/2)} \tag{8.26}$$

we obtain the following equation:

$$v^* \equiv \sqrt{\tau_w/\rho} = U_m \sqrt{f/2} \approx \sqrt{\overline{v'^2}}$$

where v^* is the *friction velocity* [m s^{-1}], which represents the order of magnitude of the velocity fluctuations in the turbulent flow near the wall. The following two important parameters for describing the velocity distribution are defined in terms of the friction velocity:

$$u^+ \equiv u/v^* \tag{8.28}$$

$$y^+ \equiv \rho v^* y/\mu \tag{8.29}$$

T. von Ka'rman [5] presented the following theoretical approach to the velocity distribution in a turbulent flow inside a circular pipe.

In the region very near to the wall, the contribution of the molecular transport mechanism is overwhelmingly dominant and we can neglect the contribution of the turbulent transport mechanism. This region is called the *laminar sub-layer*. The thickness of the laminar sub-layer is known from experiments as $y^+ = 5$. The following equation holds for the laminar sub-layer:

$$|\tau| \approx \tau_w = \mu \frac{d\bar{u}}{dy} \tag{8.30}$$

Integrating Eq. (8.30) with respect to y^+, we have the following equation:

$$u^+ = y^+ \tag{8.31}$$

In the region slightly further away from the outer edge of the laminar sub-layer ($y^+ \geq 30$), where the contribution of the turbulent transport mechanism is dominant, we can neglect the molecular transport mechanism in comparison with the turbulent one. This region is known as the *turbulent core* or fully turbulent region. From Eqs. (8.19) and (8.25), we obtain the following equation for this region:

$$\tau_w \approx \tau_{turb} = \rho l^2 \left(\frac{d\bar{u}}{dy}\right)^2 \tag{8.32}$$

From Eqs. (8.21) and (8.32), we have the following equation:

$$\frac{du^+}{dy^+} = 2.5 \frac{1}{y^+} \tag{8.33}$$

Integrating Eq. (8.33) with respect to y^+, we have the following equation:

$$u^+ = 2.5 \ln y^+ + c$$

The intermediate region between the laminar sub-layer and the turbulent core, in which the order of magnitude of the turbulent transport properties is nearly the same as that of the molecular transport properties, is usually known as the *buffer layer* ($5 \leq y^+ \leq 30$).

Better agreement with the observed data is obtained by use of the following equations:

8.2 Velocity Distribution in a Turbulent Flow inside a Circular Pipe and Friction Factors

Laminar sub-layer ($0 \leq y^+ \leq 5$):

$$u^+ = y^+ \tag{8.35a}$$

Buffer layer ($5 \leq y^+ \leq 30$):

$$u^+ = 5.0 \ln y^+ - 3.05 \tag{8.35b}$$

Turbulent core ($30 \leq y^+$):

$$u^+ = 2.5 \ln y^+ + 5.5 \tag{8.35c}$$

Equation (8.35) is referred to as von Ka'rman's *universal velocity distribution law*, and it shows fairly good agreement with observed velocity data, except for data pertaining to the region near the boundary between the buffer layer and the turbulent core. A better fitting of the data is obtained by use of van Driest's *dumping factor*, Eq. (8.22), with Eq. (8.32), or by use of the *Deissler analogy*, the details of which are discussed in the next section. Figure 8.3 shows the velocity distribution in a turbulent flow inside a smooth circular pipe. For comparison, the figure also shows the velocity distribution obtained by using *Deissler's analogy*.

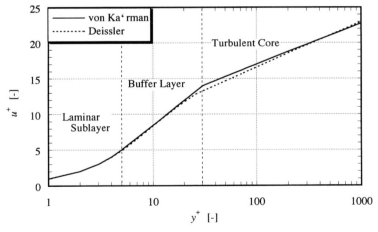

Figure 8.3 Velocity distribution in a turbulent flow inside a smooth circular pipe as obtained from von Ka'rman's universal velocity distribution law

8.2.3
Friction Factors for Turbulent Flow inside a Smooth Circular Pipe

Although Eq. (8.35c) is only valid for that part of the turbulent core that occupies the region near the wall, we will tentatively assume that it is valid even for the region near the center line of the pipe, and so we have the following equation:

$$U_{max} = v^* \left\{ 2.5 \ln \left(\frac{\rho R v^*}{\mu} \right) + 5.5 \right\} \quad (8.36)$$

Integrating Eq. (8.35c) from $y^+ = 0$ to $y^+ = R^+$ with respect to y^+, we obtain the following equation for the average velocity U_m:

$$U_m = U_{max} - 3.75 v^* \quad (8.37)$$

Substituting Eq. (8.37) into Eq. (8.36) and subsequently rearranging, we obtain the following equation for the friction factor:

$$\frac{1}{\sqrt{f}} = 4.0 \log \left(Re \sqrt{f} \right) - 0.6 \quad (8.38)$$

H. Schlichting [12] suggested replacing the second term on the right-hand side of Eq. (8.38) with 0.4, rather than the 0.6 in the original derivation, on the grounds that a better agreement with the observed data is obtained. He thus finally recommended the following equation:

$$\frac{1}{\sqrt{f}} = 4.0 \log \left(Re \sqrt{f} \right) - 0.4 \quad (8.39)$$

Predictions by Eq. (8.39) coincide with those by the well-known Blasius' empirical correlation for the Reynolds number range $5000 \leq Re \leq 10^5$:

$$f = 0.079 \, Re^{-1/4} \quad (8.40)$$

Figure 8.4 shows the friction factor in a smooth circular pipe.

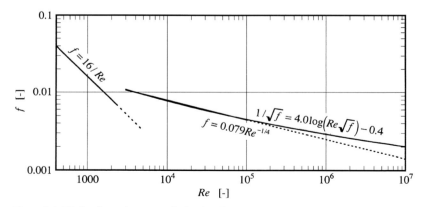

Figure 8.4 Friction factor in a smooth circular pipe

Example 8.1
Water is flowing in a smooth circular pipe of inner diameter 0.1 m. If the temperature of the water is 293 K and its average velocity is 3 m s^{-1}, calculate the thickness of the laminar sub-layer ($y^+ = 5$) and the maximum velocity of the water at the center line of the pipe.

Solution
The physical properties of water under the given conditions are as follows:

$\mu = 0.001$ Pa s, $\rho = 1000$ kg m^{-3}

Reynolds number:

$$Re = (0.1)(3.0)(1000)/(0.001) = 3.00 \times 10^5$$

Calculation of friction factor:
As the Reynolds number under the given conditions is out of range of the Blasius' correlation, we have to use Eq. (8.39), which means that we have to apply a trial-and-error calculation. As an initial value for the calculation, we will use the Blasius' correlation:

$$f_0 = (0.079)(3.00 \times 10^5)^{-0.25} = 0.00338$$

Substituting the above value into the right-hand side of Eq. (8.39), we have the first approximation f_1:

$$f_1 = \frac{1}{\{4.0 \log((3.00 \times 10^5)(0.00338)^{0.5}) - 0.4\}^2} = 0.00364$$

Repeating similar calculations:

$$f_2 = \frac{1}{\{4.0 \log((3.00 \times 10^5)(0.00364)^{0.5}) - 0.4\}^2} = 0.00362$$

$$f_3 = \frac{1}{\{4.0 \log((3.00 \times 10^5)(0.00362)^{0.5}) - 0.4\}^2} = 0.00362$$

we ultimately obtain a converged value of the friction factor

$$f = 0.00362$$

Calculation of frictional velocity:

$$v^* = (3.0)(0.00362/2)^{0.5} = 0.128 \text{ m s}^{-1}$$

The thickness of the laminar sub-layer is given by:

$$y_{sub} = (5)(0.001)/(0.128)(1000) = 3.9 \times 10^{-5} \text{ m}$$

which indicates that only 0.08% of the cross-section is occupied by the laminar sub-layer.

The maximum velocity at the center line of the pipe can be estimated by means of Eq. (8.35c):

$$R^+ = (0.05)(0.128)(1000)/(0.001) = 6.4 \times 10^3$$

$$U^+_{max} = 2.5 \ln(6.4 \times 10^3) + 5.5 = 27.4$$

$$U_{max} = (27.4)(0.128) = 3.51 \text{ m s}^{-1}$$

$$U_m/U_{max} = (3.0)/(3.51) = 0.857$$

On the other hand, the maximum velocity can also be calculated by the use of 1/7-th power law, Eq. (8.23):

$$\frac{U_m}{U_{max}} = \frac{(2)(7)^2}{(7+1)\{(2)(7)+1\}} = 0.816$$

$$U_{max} = (3.0)/(0.816) = 3.68 \text{ m s}^{-1}$$

For the maximum velocity at the center of the pipe, Eq. (8.35c) gives a 5% smaller value than that obtained by the *1/7-th power law*. This may be due to the fact that Eq. (8.35c) is only valid for the part of the turbulent core that is near to the wall region.

8.3
Analogy between Momentum, Heat, and Mass Transfer

Transport phenomena in turbulent flow are much more complicated than those in laminar flow and we cannot obtain any analytical solution of the time-averaged governing equations, not even in the simplest case. Discussions in the previous sections may suggest that there is some kind of mathematical similarity between momentum, heat, and mass transfer:

$$\tau = -\rho(\nu + \varepsilon)\frac{d\bar{u}}{dy} \tag{8.11}$$

$$q = -\rho c_p(\alpha + \varepsilon_H)\frac{d\bar{T}}{dy} \tag{8.13}$$

$$J = -\rho(D + \varepsilon_D)\frac{d\overline{\omega}}{dy} \tag{8.15}$$

and this fact suggests another possible means of estimating heat and mass transfer rates in a turbulent flow, which, although approximate, is practically useful. The similarity of the transport equations relating to momentum, heat, and mass transfer suggests that the friction factors and the heat or diffusional fluxes at the wall are mutually interrelated, and so if any one of the above quantities is known by some means, then the other two quantities may be estimated by assuming mutual relationships. The problem then reduces to finding mutual relationships between the friction factors and the heat or diffusional fluxes or that between the heat and diffusional fluxes. This indirect method of predicting transport phenomena in turbulent flow is known as *analogy*.

8.3.1
Reynolds Analogy

In the turbulent core in the region near the wall, the contribution of the turbulent transport mechanism far outweighs that of the molecular mechanism. O. Reynolds was the first to apply the concept of analogy to the prediction of turbulent transport phenomena by assuming the following equation:

$$\varepsilon \approx \varepsilon_H \approx \varepsilon_D \tag{8.41}$$

Rearranging Eqs. (8.11) and (8.13) under the above assumption, we obtain the following equation:

$$\frac{|\tau|}{|q|} = \frac{1}{c_p}\frac{d\overline{u}}{d\overline{T}} \tag{8.42}$$

Although Eq. (8.42) is only valid for the turbulent core in the region near the wall, integrating this equation from the wall to the center of the pipe with respect to y^+, the following equation is obtained:

$$\frac{|\tau_w|}{|q_w|} = \frac{1}{c_p}\frac{U_m}{|T_s - \overline{T_\infty}|} \tag{8.43}$$

Rearranging Eq. (8.43) by use of the definition of the friction factor, we have:

$$\frac{f}{2} \equiv \frac{|\tau_w|}{\rho U_m^2} = \frac{|q_w|}{\rho c_p U_m |T_s - \overline{T_\infty}|}$$

$$= \left(\frac{|q_w|D_T}{\kappa|T_s - \overline{T_\infty}|}\right)\left(\frac{\mu}{\rho U_m D_T}\right)\left(\frac{\kappa}{c_p \mu}\right) = \frac{Nu}{RePr} \equiv St_H \tag{8.44}$$

In a similar way, the following equation is obtained by rearrangement of Eqs. (8.11) and (8.15):

$$\frac{f}{2} = \frac{Sh(1-\omega_s)}{ReSc} \equiv St_M(1-\omega_s) \tag{8.45}$$

where the term in parentheses on the right-hand side of Eq. (8.45) is due to the assumption of unidirectional diffusion. Equations (8.44) and (8.45) are referred to as the *Reynolds analogy* in honor of O. Reynolds. The significance of the Reynolds analogy is that it was the first approach to describing the turbulent transport mechanism from the view point of analogy, although it is only valid for the special case of $Pr = Sc = 1$.

8.3.2
Chilton–Colburn Analogy

T. H. Chilton and A. P. Colburn [1] proposed a well-known correlation on the basis of wide ranges of heat and mass transfer data, as shown below:

$$\frac{f}{2} = j_H = j_D \tag{8.46}$$

where j_H and j_D are the *j-factors* for heat and mass transfer, respectively, which are defined by the following equations:

$$j_H \equiv St_H Pr^{2/3} = \frac{Nu}{RePr^{1/3}} \tag{8.47}$$

$$j_D \equiv St_M(1-\omega_s)Sc^{2/3} = \frac{Sh(1-\omega_s)}{ReSc^{1/3}} \tag{8.48}$$

Equations (8.47) and (8.48) are usually referred to as the *Chilton–Colburn analogy*.

Although they are empirical correlations, the same equations can be obtained in a semi-theoretical manner by considering the transport mechanism in the laminar sub-layer. Therein, the contribution of the molecular transport mechanism far outweighs that of the turbulent mechanism, and we can approximate the transport phenomena in this region by neglecting the turbulent transport properties.

Rearranging the definition of the friction factor, we obtain the following equation:

$$\frac{f}{2} = \frac{|\tau_w|}{\rho U_m^2} = \frac{1}{Re}\left\{\frac{d(\bar{u}/U_m)}{d\eta}\right\}_{wall}\left(\frac{D_T}{\delta}\right) \tag{8.49}$$

8.3 Analogy between Momentum, Heat, and Mass Transfer

where D_T is the inner diameter of the circular pipe [m], U_m is the average velocity [m s^{-1}], δ is the thickness of the laminar sub-layer [m], and $\eta\,(\equiv y/\delta)$ is the dimensionless distance from the wall.

On the other hand, we have the following equation from the definition of the Sherwood number:

$$Sh(1-\omega_s) = \frac{N_A(1-\omega_s)}{\rho D(\omega_s - \overline{\omega}_{av})} = \left(\frac{d\theta_c}{d\eta}\right)\left(\frac{D_T}{\delta}\right) \qquad (8.50)$$

where θ_c is the dimensionless concentration defined by the following equation:

$$\theta_c \equiv \frac{(\omega_s - \overline{\omega})}{(\omega_s - \overline{\omega}_{av})} \qquad (8.51)$$

Substituting Eq. (8.50) into Eq. (8.49) and rearranging, the following equation is obtained:

$$\frac{f}{2} = \frac{Sh(1-\omega_s)}{Re} \frac{\left\{\dfrac{d(\bar{u}/U_m)}{d\eta}\right\}_{wall}}{\left(\dfrac{d\theta_c}{d\eta}\right)_{wall}} \approx \frac{Sh(1-\omega_s)}{Re}\left(\frac{\delta_c}{\delta}\right) \qquad (8.52)$$

Substituting Eq. (3.35), the equation for the thickness of the concentration boundary layer:

$$(\delta_c/\delta) \approx Sc^{-1/3} \qquad (3.35)$$

into Eq. (8.52), we obtain the following equation:

$$\frac{f}{2} = \frac{Sh(1-\omega_s)}{Re\,Sc^{1/3}} = j_D \qquad (8.53)$$

The same equation as Eq. (8.48) is obtained. In a similar way, we can also derive Eq. (8.47).

Figure 8.5 shows a comparison of the Chilton–Colburn analogy with experimental data on the evaporation of various liquids into air using a wetted-wall column obtained by E. R. Gilliland and T. K. Sherwood [4]. The observed data show nearly the same tendency as the prediction by Eq. (8.53), but the values are about 10–20% higher than predicted.

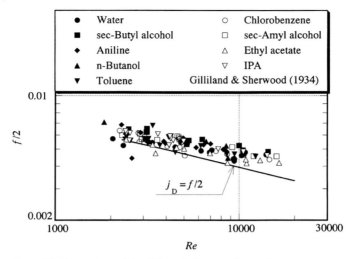

Figure 8.5 Comparison of the Chilton–Colburn analogy with experimental data obtained for a wetted-wall column by E. R. Gilliland and T. K. Sherwood [4]

Example 8.2

The drag coefficients of a spherical particle can be approximated by the following equation for the Reynolds number range $20 \leq Re_p \leq 300$:

$$C_D = 11\, Re_p^{-1/2}$$

Derive the equation for the Sherwood number as a function of the Reynolds and Schmidt numbers by applying the Chilton–Colburn analogy if the relative order of magnitude of the frictional drag compared to the total drag is 1/2 for the aforementioned Reynolds number range.

Solution

$$F_{vis} = \frac{1}{2} F_D = \frac{c_D}{2}\left(\frac{\pi D_p^2}{4}\right)\left(\frac{\rho U^2}{2}\right) = \left(\frac{11}{16}\right)\left(\frac{\pi D_p^2 \rho U^2}{Re_p^{1/2}}\right) \quad \text{(A)}$$

$$\tau_{vis} = \left(\frac{F_{vis}}{\pi D_p^2}\right) = \frac{11}{16}\left(\frac{\rho U^2}{Re_p^{1/2}}\right) \quad \text{(B)}$$

Substituting Eq. (B) into the definition of the friction factor, we have:

$$f \equiv \frac{\tau_{vis}}{(\rho U^2/2)} = \frac{11}{8}\, Re_p^{-1/2} \quad \text{(C)}$$

From the Chilton–Colburn analogy

$$Sh(1 - \omega_s) = \frac{f}{2} Re_p Sc^{1/3} \tag{D}$$

Substituting Eq. (C) into Eq. (D), we obtain the following equation:

$$Sh(1 - \omega_s) = \frac{11}{16} Re_p^{1/2} Sc^{1/3} = 0.68 Re_p^{1/2} Sc^{1/3} \tag{E}$$

The right-hand side of the above equation has nearly the same form as the second term of the Ranz–Marshall correlation ($0.6\, Re_p^{1/2} Sc^{1/3}$), with only a slight difference in the numerical coefficient.

8.3.3
Von Ka'rman Analogy

The Reynolds analogy and the Chilton–Colburn analogy are based on the transport mechanisms for the two extreme cases, the turbulent core and the laminar sub-layer. T. von Ka'rman [5] proposed the following equations by considering the contribution of the transport mechanisms in all three of the regions, with the assumption that the velocity, temperature, and concentration distributions are similar in each region:

For heat transfer:

$$St_H = \frac{\frac{1}{2}f}{1 + 5\sqrt{\frac{f}{2}}\left[Pr - 1 + \ln\left\{1 + \frac{5}{6}(Pr - 1)\right\}\right]} \tag{8.54}$$

For mass transfer:

$$St_M(1 - \omega_s) = \frac{\frac{1}{2}f}{1 + 5\sqrt{\frac{f}{2}}\left[Sc - 1 + \ln\left\{1 + \frac{5}{6}(Sc - 1)\right\}\right]} \tag{8.55}$$

Equations (8.54) and (8.55) are referred to as the *von Ka'rman analogy*. For large Schmidt or Prandtl number ranges, Eqs. (8.54) and (8.55) reduce to the following equations:

$$Nu = 0.04\, Re^{7/8} \tag{8.56}$$

$$Sh(1 - \omega_s) = 0.04\, Re^{7/8} \tag{8.57}$$

which may lead to the unrealistic conclusion that the Nusselt or Sherwood numbers are independent of the Prandtl or Schmidt numbers. For this reason, the von Ka'rman analogy is not valid for liquid-phase mass transfer, where Schmidt numbers are very large.

8.3.4
Deissler Analogy

R. G. Deissler [2] proposed the following semi-theoretical equation for the eddy kinematic viscosities in the region near the wall on the basis of his measurements of heat transfer inside a circular pipe:

$$y^+ < 26$$

$$\frac{\varepsilon}{\nu} = n^2 u^+ y^+ \{1 - \exp(-n^2 u^+ y^+)\} \tag{8.58}$$

For the numerical value of the coefficient n, he recommends $n = 0.124$. Substituting Eq. (8.58) into Eq. (8.11) and integrating with respect to y^+, the following equation is obtained:

$$u^+ = \int_0^{y^+} \frac{dy^+}{1 + n^2 u^+ y^+ \{1 - \exp(-n^2 u^+ y^+)\}} \tag{8.59}$$

Assuming the following relationship for the eddy transport properties:

$$\varepsilon \approx \varepsilon_H \approx \varepsilon_D \tag{8.41}$$

and integrating Eqs. (8.13) and (8.15) with respect to y^+, similar equations for the distributions of the dimensionless temperature and the concentration are obtained:

Temperature:

$$T^+ \equiv \frac{\rho c_p (T_w - \overline{T}) U_m \sqrt{f/2}}{q_w} = \int_0^{y^+} \frac{dy^+}{(1/Pr) + n^2 u^+ y^+ \{1 - \exp(-n^2 u^+ y^+)\}} \tag{8.60}$$

Concentration:

$$C^+ \equiv \frac{(\omega_s - \overline{\omega}) U_m \sqrt{f/2}}{J_w} = \int_0^{y^+} \frac{dy^+}{(1/Sc) + n^2 u^+ y^+ \{1 - \exp(-n^2 u^+ y^+)\}} \tag{8.61}$$

In the part of the turbulent core region slightly further away from the wall ($y^+ \geq 26$), Deissler assumed a logarithmic velocity distribution law of the form:

$$u^+ - u_1^+ = \frac{1}{0.36}\ln\left(\frac{y^+}{26}\right) \tag{8.62}$$

where u_1^+ is the numerical value of u^+ at $y^+ = 26$.

Similar equations are assumed for the temperature and concentration distributions in this region:

$$T^+ = \frac{1}{0.36}\ln\left(\frac{y^+}{26}\right) + T_1^+ \tag{8.63}$$

$$C^+ = \frac{1}{0.36}\ln\left(\frac{y^+}{26}\right) + C_1^+ \tag{8.64}$$

where T_1^+ and C_1^+ are the numerical values of T^+ and C^+, respectively, at $y^+ = 26$.

The Stanton number for heat and mass transfer can easily be calculated from the average temperature and concentration as shown below:

Heat transfer:

$$T_b^+ \equiv \frac{\int_0^{R^+} T^+ u^+ (R^+ - y^+) dy^+}{\int_0^{R^+} u^+ (R^+ - y^+) dy^+} = \frac{\sqrt{f/2}}{St_H} \tag{8.65}$$

Mass transfer:

$$C_b^+ \equiv \frac{\int_0^{R^+} C^+ u^+ (R^+ - y^+) dy^+}{\int_0^{R^+} u^+ (R^+ - y^+) dy^+} = \frac{\sqrt{f/2}}{St_M(1-\omega_s)} \tag{8.66}$$

Equations (8.65) and (8.66) reduce to much simpler forms for large Prandtl or Schmidt numbers, as shown below:

$Pr \gg 1$
$$St_H = 0.079\sqrt{f}\, Pr^{-3/4} \tag{8.67}$$
$Sc \gg 1$

$$St_M(1-\omega_s) = 0.079\sqrt{f}\, Sc^{-3/4} \tag{8.68}$$

Figure 8.6 shows a comparison of the four analogy models relating to turbulent flow, the Reynolds, the Chilton–Colburn, the von Ka'rman, and the Deissler analogies. In the figure, the predictions by the von Ka'rman and Deissler analogies are

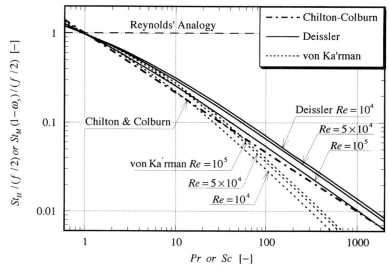

Figure 8.6 Comparison of various analogy models

seen to show a slight dependence on the Reynolds number. The Deissler analogy is the best among the four analogy models, in that it agrees very well with the observed data for a wide range of Prandtl and Schmidt numbers and can even be used to predict the velocity, temperature, and concentration distributions, although the calculations are rather complicated and elaborate. For mass transfer, however, the Chilton–Colburn analogy shows fairly good agreement with the observed data over a wide range of Schmidt numbers, in spite of its simple mathematical form. An important fact about the Chilton–Colburn analogy is that it is not only valid for turbulent flow in a circular pipe but also for various other types of turbulent flow.

Example 8.3

Butanol at 323 K is counter-currently contacted with air at 109.3 kPa and 323 K in a wetted-wall column of inner diameter 0.0267 m and length 1.17 m. Calculate the rate of evaporation of the butanol [kg s^{-1}] if the flow rate of the air is 0.100 kg min^{-1}.

Solution
Step 1. Estimation of physical properties:

Assuming that the surface temperature of the butanol is 323 K, the physical properties of the system are as follows:

$\rho = 1.18$ kg m^{-3}, $\mu = 1.95 \times 10^{-5}$ Pa s, $Sc = 1.87$, $p_s = 5.10$ kPa

8.3 Analogy between Momentum, Heat, and Mass Transfer

Step 2. Calculation of the friction factor:

$$U_m = \frac{(0.100)}{(60)(0.785)(0.0267)^2(1.18)} = 2.52 \text{ m s}^{-1}$$

$$Re = \frac{(1.18)(0.0267)(2.52)}{(1.95 \times 10^{-5})} = 4072$$

From Blasius' correlation:

$$f = (0.079)(4072)^{-0.25} = 9.89 \times 10^{-3}$$

$$v^* = (2.52)\sqrt{(9.89 \times 10^{-3})/(2)} = 0.177 \text{ m s}^{-1}$$

$$R^+ = \frac{(1.18)(0.0267/2)(0.177)}{(1.95 \times 10^{-5})} = 143$$

Step 3. Calculation of the Stanton number for mass transfer by the use of various analogy models:

a) Chilton–Colburn analogy

$$St_M(1 - \omega_s) = \frac{(0.00989/2)}{(1.87)^{2/3}} = 3.25 \times 10^{-3}$$

b) von Ka'rman analogy

$$St_M(1 - \omega_s) = \frac{(0.00989/2)}{1 + (5)\sqrt{\left(\frac{0.00989}{2}\right)}\left[1.87 - 1 + \ln\left\{1 + \frac{5}{6}(1.87 - 1)\right\}\right]}$$

$$= 3.30 \times 10^{-3}$$

c) Deissler analogy

Table 8.1 shows the results of numerical calculations by Eqs. (8.59)–(8.66). In the table, the second column represents the dimensionless velocity distribution calculated by Eqs. (8.59) and (8.62) and the fifth column the dimensionless concentration calculated by Eqs. (8.61) and (8.64). The seventh column represents numerical integration of $u^+ C^+ (R^+ - y^+)$ with respect to y^+ for the interval between $y^+ = 0$ and $y^+ = y^+$. Figure 8.7 shows the velocity and the concentration distribution calculated by the Deissler analogy.

Table 8.1 Calculation of the rate of mass transfer, the velocity, and the concentration distribution in a wetted-wall column by use of the Deissler analogy in Example 8.3

y^+	u^+	$u^+(R^+-y^+)$	$\int_0^{y^+} u^+(R^+-y^+)dy^+$	C^+	$u^+C^+(R^+-y^+)$	$\int_0^{y^+} u^+C^+(R^+-y^+)dy^+$
0	0.0	0.00E+00	0.00E+00	0.0	0.00E+00	0.00E+00
1	1.0	1.42E+02	7.10E+01	0.9	1.33E+02	6.65E+01
2	2.0	2.82E+02	2.83E+02	2.8	7.91E+02	5.29E+02
3	3.0	4.19E+02	6.33E+02	4.6	1.94E+03	1.90E+03
4	4.0	5.49E+02	1.12E+03	6.4	3.52E+03	4.63E+03
5	4.9	6.73E+02	1.73E+03	8.0	5.41E+03	9.09E+03
6	5.7	7.85E+02	2.46E+03	9.5	7.43E+03	1.55E+04
7	6.5	8.85E+02	3.29E+03	10.7	9.47E+03	2.39E+04
8	7.2	9.73E+02	4.22E+03	11.7	1.14E+04	3.44E+04
9	7.8	1.05E+03	5.23E+03	12.6	1.32E+04	4.67E+04
10	8.4	1.11E+03	6.32E+03	13.3	1.49E+04	6.08E+04
12	9.3	1.22E+03	8.65E+03	14.5	1.77E+04	9.34E+04
14	10.1	1.30E+03	1.12E+04	15.5	2.01E+04	1.31E+05
16	10.7	1.36E+03	1.38E+04	16.2	2.20E+04	1.73E+05
18	11.2	1.41E+03	1.66E+04	16.8	2.36E+04	2.19E+05
20	11.7	1.44E+03	1.94E+04	17.4	2.50E+04	2.68E+05
22	12.1	1.47E+03	2.23E+04	17.8	2.62E+04	3.19E+05
24	12.5	1.49E+03	2.53E+04	18.2	2.72E+04	3.72E+05
26	12.9	1.50E+03	2.83E+04	18.6	2.80E+04	4.27E+05
30	13.3	1.50E+03	3.43E+04	19.0	2.85E+04	5.40E+05
40	14.1	1.45E+03	4.90E+04	19.8	2.87E+04	8.26E+05
50	14.7	1.36E+03	6.31E+04	20.4	2.79E+04	1.11E+06
60	15.2	1.26E+03	7.62E+04	20.9	2.64E+04	1.38E+06
80	16.0	1.01E+03	9.88E+04	21.7	2.19E+04	1.86E+06
100	16.6	7.13E+02	1.16E+05	22.4	1.59E+04	2.24E+06
120	17.1	3.93E+02	1.27E+05	22.9	8.99E+03	2.49E+06
140	17.5	5.26E+01	1.32E+05	23.3	1.22E+03	2.59E+06
143	17.9	0.00E+00	1.32E+05	23.4	0.00E+00	2.59E+06

From Tab. 8.1 and Eq. (8.66):

$$C_b^+ = (2.59 \times 10^6)/(1.32 \times 10^5) = 19.7$$

$$St_M(1 - \omega_s) = \frac{\sqrt{(0.00989)/2}}{19.7} = 3.57 \times 10^{-3}$$

Step 4. Calculation of the rate of evaporation:

Surface area for evaporation:

$$S = (3.14)(0.0267)(1.17) = 9.81 \times 10^{-2} \text{ m}^2$$

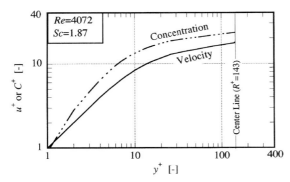

Figure 8.7 The velocity and concentration distributions in Example 8.3 calculated by the Deissler analogy

Surface concentration and concentration driving force:

$$\omega_s = \frac{\{(5.10)/(109.3)\}(74)}{\{1-(5.10)/(109.3)\}(29)+\{(5.10)/(109.3)\}(74)} = 0.1111$$

$$\Delta\overline{\omega} = \frac{(0.1111)+(0.1111-\overline{\omega_2})}{2} = 0.1111 - 0.5\,\overline{\omega_2}$$

Rate of evaporation: From the material balance, we have the following equation between the rate of evaporation W [kg s^{-1}] and the vapor concentration at the top of the column $\overline{\omega_2}$:

$$W = N_A S = \frac{St_M(1-\omega_s)(1.18)(2.52)(9.81\times10^{-2})}{(1-0.1111)}(0.1111-0.5\,\overline{\omega_2})$$

$$= C(0.1111 - 0.5\,\overline{\omega_2}) \tag{A-1}$$

$$\overline{\omega_2} = \frac{W}{(0.100)/(60)+W} \tag{A-2}$$

Using the above equations, the rate of evaporation is calculated by a *trial-and-error* method. The results are shown in Tab. 8.2.

Table 8.2 Comparison of the various analogy models in the case of Example 8.3

	Chilton & Colburn	von Ka'rman	Deissler	Observed value
$St_M(1-\omega_s)$	3.25E-03	3.30E-03	3.57E-03	
Coefficient C	1.07E-03	1.08E-03	1.17E-03	
Top vapor concentration $\overline{\omega_2}$ [–]	0.0475	0.0491	0.0508	
Evaporation rate W [kg s^{-1}]	8.30E-05	8.58E-05	8.90E-05	9.27E-05

The experimental rate of evaporation corresponding to the specified conditions has been reported as $W = 5.56$ g min^{-1} (= 9.27×10^{-5} kg s^{-1}) by Gilliland and Sherwood [4].

8.4
Friction Factor, Heat, and Mass Transfer in a Turbulent Boundary Layer

8.4.1
Velocity Distribution in a Turbulent Boundary Layer

Experimental results relating to a turbulent boundary layer along a flat plate at zero incidence indicate that the velocity distribution can be expressed either by the *logarithmic velocity distribution law*:

$$u^+ = 2.5 \ln y^+ + 5.5 \tag{8.35c}$$

or by the *1/n-th power law*:

$$\left(\frac{\bar{u}}{U}\right) = \left(\frac{y}{\delta}\right)^{1/n} \tag{8.69}$$

where δ is the thickness of the turbulent boundary layer [m].

Important parameters, such as the *displacement thickness*, δ_1, and the *momentum thickness*, δ_2, which are defined by the following equations, can easily be calculated if the velocity distribution in the boundary layer follows the 1/n-th power law.

Displacement thickness:

$$\left(\frac{\delta_1}{\delta}\right) = \frac{1}{\delta}\int_0^\delta \left(1 - \frac{u}{U}\right)dy = \frac{1}{\delta}\int_0^\delta \left\{1 - \left(\frac{y}{\delta}\right)^{1/n}\right\}dy = \frac{n}{(n+1)} \tag{8.70}$$

Momentum thickness:

$$\left(\frac{\delta_2}{\delta}\right) \equiv \frac{1}{\delta}\int_0^\delta \left(\frac{u}{U}\right)\left\{1 - \left(\frac{u}{U}\right)\right\}dy$$

$$= \frac{1}{\delta}\int_0^\delta \left(\frac{y}{\delta}\right)^{1/n}\left\{1 - \left(\frac{y}{\delta}\right)^{1/n}\right\}dy = \frac{n}{(n+1)(n+2)} \tag{8.71}$$

8.4.2
Friction Factor

Figure 8.8 shows a schematic representation of momentum transfer in a turbulent boundary layer along a flat plate. The time-averaged two-dimensional equation of motion can be written as follows:

$$\bar{u}\frac{\partial \bar{u}}{\partial x} + \bar{v}\frac{\partial \bar{u}}{\partial x} = \frac{1}{\rho}\frac{\partial \tau}{\partial y} \qquad (8.72)$$

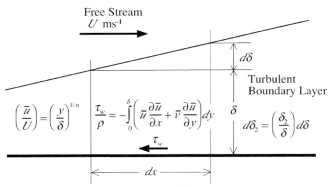

Figure 8.8 Momentum transfer in a turbulent boundary layer

Integrating Eq. (8.72) with respect to y from the wall to the outer edge of the boundary layer, δ, we obtain the following equation:

$$\frac{\tau_w}{\rho} = -\int_0^\delta \left(\bar{u}\frac{\partial \bar{u}}{\partial x} + \bar{v}\frac{\partial \bar{u}}{\partial y}\right)dy \qquad (8.73)$$

From the time-averaged continuity equation, we have:

$$\bar{v} = -\int_0^y \frac{\partial \bar{u}}{\partial x}dy \qquad (8.74)$$

Substituting Eq. (8.74) into Eq. (8.73) and integrating the second term in parentheses on the right-hand side of the above equation by parts, we obtain the following equation:

$$\frac{\tau_w}{\rho U^2} = -\frac{1}{U^2}\int_0^\delta \left(\bar{u}\frac{\partial \bar{u}}{\partial x} - \frac{\partial \bar{u}}{\partial y}\int_0^\delta \frac{\partial \bar{u}}{\partial x}dy\right)dy = \frac{d}{dx}\int_0^\delta \left(\frac{\bar{u}}{U}\right)\left(1 - \frac{\bar{u}}{U}\right)dy \qquad (8.75)$$

Rearranging Eq. (8.75) by use of the definition of the friction factor, we have the following equation:

$$\frac{f_x}{2} \equiv \frac{\tau_w}{\rho U^2} = \frac{d\delta_2}{dx} \tag{8.76}$$

H. Schlichting [12] derived the following equation for the velocity distribution in a turbulent boundary layer from the 1/7-th power law:

$$\left(\frac{\bar{u}}{v^*}\right) = 8.74 \left(\frac{\rho v^* y}{\mu}\right)^{1/7} \tag{8.77}$$

He also derived the following equation for the thickness of the turbulent boundary layer from Eqs. (8.26) and (8.77):

$$\delta = 0.37 x \left(\frac{\rho U x}{\mu}\right)^{-0.2} \tag{8.78}$$

Substituting Eqs. (8.71) and (8.78) into Eq. (8.76), the following equation for the local friction factor in a turbulent boundary layer can be obtained:

$$f_x = 0.0576 \, Re_x^{-0.2} \tag{8.79}$$

where Re_x is the local Reynolds number defined by Eq. (8.80).

$$Re_x \equiv \rho U x / \mu \tag{8.80}$$

Schlichting [12] suggests that better agreement with observed data is obtained if the numerical value of the coefficient on the right-hand side of Eq. (8.79) is replaced with 0.0592 instead of 0.0576 in the original derivation. The recommended equation for the local friction factor is thus:

$$f_x = 0.0592 \, Re_x^{-0.2} \tag{8.81)]}$$

The average friction factor over the entire length of the plate, L, can easily be obtained by integrating Eq. (8.81) with respect to x by the following equation:

$$f \equiv \frac{\overline{\tau_w}}{(\rho U^2/2)} = \frac{1}{L} \int_0^L f_x dx = 0.074 \, Re^{-0.2} \tag{8.82}$$

where is the Reynolds number over the total length of the plate:

$$Re \equiv \rho U L / \mu \tag{8.83}$$

8.4.3
Heat and Mass Transfer in a Turbulent Boundary Layer

From the Chilton–Colburn analogy together with the equation for the local friction factor, Eq. (8.81), we can obtain the following equations for heat and mass transfer for a turbulent flow along a flat plate:

$$St_{Hx} = 0.0296 \, Pr^{-2/3} \, Re_x^{-0.2} \tag{8.84}$$

$$St_{Mx}(1 - \omega_s) = 0.0296 \, Sc^{-2/3} \, Re_x^{-0.2} \tag{8.85}$$

R. J. Moffat and W. M. Kays [8] described an experimental approach to heat transfer in a turbulent flow along a porous flat plate and proposed the following equation:

$$St_{Hx} = 0.0292 \, Pr^{-0.4} \, Re_x^{-0.2} \tag{8.86}$$

For mass transfer, we can theoretically derive a similar equation to Eqs. (8.84)–(8.86) from the time-averaged two-dimensional diffusion equation. Integrating the diffusion equation with respect to y from the wall to the outer edge of the concentration boundary layer, δ_c, we obtain the following equation:

$$St_{Mx}(1 - \omega_s) = \frac{d}{dx} \int_0^{\delta_c} \left(\frac{u}{U}\right)\left(1 - \frac{\omega_s - \overline{\omega}}{\omega_s - \overline{\omega}_\infty}\right) dy \tag{8.87}$$

In the following, we assume the case of large Schmidt number and that the concentration distribution follows the 1/7-th power law:

$$\theta_c \equiv \frac{\omega_s - \overline{\omega}}{\omega_s - \overline{\omega}_\infty} = \left(\frac{y}{\delta_c}\right)^{1/7} \tag{8.88}$$

Since we have assumed the case of large Schmidt number, the thickness of the concentration boundary layer is very small in comparison with that of the velocity boundary layer, and so the velocity distribution inside the concentration boundary layer can be approximated by the following equation:

$$\frac{\overline{u}}{U} \approx \left(\frac{y}{\delta}\right) \tag{8.89}$$

Substituting Eqs. (8.88) and (8.89) into Eq. (8.87), we obtain the following equation:

$$St_{Mx}(1 - \omega_s) \approx \frac{d}{dx} \int_0^{\delta_c} \left(\frac{y}{\delta}\right)\left\{1 - \left(\frac{y}{\delta_c}\right)^{1/7}\right\} dy = \frac{1}{30}\left(\frac{\delta_c}{\delta}\right)^2 \frac{d\delta}{dx} \tag{8.90}$$

Substituting Eq. (8.78) into Eq. (8.90) and rearranging by consideration of Eq. (3.35), we finally obtain the following equation for mass transfer in a turbulent boundary layer along a flat plate:

$$St_{Mx}(1-\omega) \approx 0.01\, Sc^{-2/3} Re_x^{-0.2} \tag{8.91}$$

An equation similar to Eq. (8.86) can thus be obtained theoretically. The sole difference is the numerical coefficient on the right-hand side of the equation.

8.5
Turbulent Boundary Layer with Surface Mass Injection or Suction

In high-temperature operations, a small amount of cold gas is injected into the high-temperature free stream gas, or a small amount of liquid is injected and forced to evaporate through the porous wall, this being a useful technique for protecting the wall from the high-temperature gas. Under these circumstances, the high mass flux effect will be operative and Mickley's film model described in Section 4.8.2 will apply for prediction of the friction factors and the heat and mass transfer rates.

Effect of mass injection or suction on the friction factor:

$$\frac{f_x}{f_{x0}} = \frac{b}{(e^b - 1)} = \frac{\ln(1 + B_u)}{B_u} \tag{4.95}$$

$$B_u \equiv \left(\frac{v_w}{U}\right)\left(\frac{2}{f_x}\right) = b\left(\frac{f_{x0}}{f_x}\right) \tag{4.97}$$

where f_{x0} is the local friction factor without mass injection or suction, f_x is the local friction factor, and B_u and b are the blowing parameters defined by Eq. (4.97). Equation (4.95) shows good agreement with observed data by H. S. Mickley et al. [7] and with that obtained by W. M. Kays [6].

Effect of mass injection or suction on heat transfer:

$$\frac{St_{Hx}}{St_{Hx0}} = \frac{b_H}{(e^{b_H} - 1)} = \frac{\ln(1 + B_H)}{B_H} \tag{4.98}$$

$$B_H = \left(\frac{v_w}{U}\right)\left(\frac{1}{St_{Hx}}\right) = b_H\left(\frac{St_{Hx0}}{St_{Hx}}\right) \tag{4.100}$$

where $St_{Hx} (\equiv Nu_x/Re_x Pr)$ and $St_{Hx0} (\equiv Nu_{x0}/Re_x Pr)$ are the local Stanton numbers for heat transfer with and without mass injection or suction, and B_H and b_H are the transfer number for heat transfer and the blowing parameter defined by Eq. (4.100). Figure 8.9 shows a comparison of Mickley's film model for the effect on heat transfer of mass injection or suction in a turbulent boundary layer along a

8.5 Turbulent Boundary Layer with Surface Mass Injection or Suction

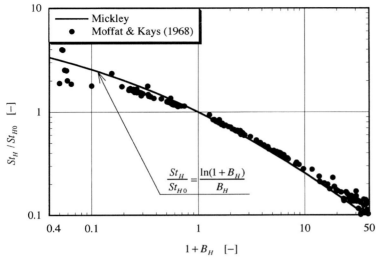

Figure 8.9 Effect of mass injection or suction on heat transfer in a turbulent boundary layer; comparison of Mickley's film model with observed data by R. J. Moffat and W. M. Kays [8]

flat plate with observed data by R. J. Moffat and W. M. Kays [8]; good agreement is evident.

Effect of mass injection or suction on mass transfer:

$$\frac{St_{Mx}}{St_{Mx0}} = \frac{b_M}{(e^{b_M} - 1)} = \frac{\ln(1 + B_M)}{B_M} \tag{4.99}$$

$$B_M = \left(\frac{v_w}{U}\right) \frac{1}{\{St_{Mx}(1 - \omega_s)\}} = b_M \frac{St_{Mx0}}{St_{Mx}} \tag{4.101}$$

where $St_{Mx} (\equiv Sh_x/Re_x Sc)$ and $St_{Mx0} (\equiv Sh_{x0}/Re_x Sc)$ are the local Stanton numbers for mass transfer with and without mass injection or suction, and B_M and b_M are the transfer number and blowing parameter for mass transfer defined by Eq. (4.101). Although Eq. (4.99) may seem to hold for mass transfer in a turbulent boundary layer, no experimental approach has ever been attempted.

Example 8.4

Air at 333 K and 1 atm is flowing along a flat porous plate of length 1 m and width 0.2 m at a velocity of 15 m s^{-1}. The surface temperature of the porous plate is 293 K and air at 293 K is injected from the surface of the porous plate into the free stream air at a velocity that is 0.2% of the free stream velocity. Calculate the average heat flux over the porous plate. The local heat flux of the plate without mass injection is given, according to Moffat and Kays [8], by the following equation:

$$St_{Hx0} = 0.0292\, Pr^{-0.4}\, Re_x^{-0.2}$$

Solution

The physical properties of the system under the specified conditions are as follows:

$\rho_{333K} = 1.07$ kg m^{-3}, $\mu_{333K} = 1.95 \times 10^{-5}$ Pa s, $c_{p333K} = 1.01 \times 10^3$ J kg^{-1} K^{-1}

$Pr_{air} = 0.71$

The average Stanton number for heat transfer is obtained from the local flux correlation by integrating with respect to x:

$$\overline{(St_H)}_0 = \frac{1}{L}\int_0^L (0.0292\, Pr^{-0.4}\, Re_x^{-0.2})\, dx = 0.0365\, Pr^{-0.4}\, Re_x^{-0.2}$$

$Re = (1.07)(15)(1.0)/(1.95 \times 10^{-5}) = 8.23 \times 10^5$

The average Stanton number for heat transfer without mass injection is given by:

$$\overline{(St_H)}_0 = (0.0365)(0.71)^{-0.4}(8.23 \times 10^5)^{-0.2} = 2.75 \times 10^{-3}$$

The blowing parameter for heat transfer, b_H, is given by:

$$b_H = (0.002)/(2.75 \times 10^{-3}) = 0.723$$

From Eq. (4.98), we have the following value:

$$\frac{\overline{St_H}}{\overline{St_{H0}}} = \frac{(0.723)}{\exp(0.723) - 1} = 0.682$$

which indicates that the heat flux will reduce to 68% as a result of 0.2% mass injection.

$$\overline{St_H} = (2.75 \times 10^{-3})(0.682) = 1.88 \times 10^{-3}$$

The average heat flux over the plate under the specified conditions is given by the following equation:

$$q_w = (1.88 \times 10^{-3})(1.07)(1.01 \times 10^3)(15)(333 - 293) = 1.22 \text{ kW m}^{-2}$$

References

1 T. H. Chilton and A. P. Colburn, „Mass Transfer (Absorption) Coefficients: Prediction from Data on Heat Transfer and Fluid Friction", *Industrial and Engineering Chemistry*, **26**, 1183–1187 (1934).

2 R. G. Deissler, „Analysis of Turbulent Heat Transfer, Mass Transfer, and Friction in Smooth Tubes at High Prandtl and Schmidt Numbers", *NACA Reports*, **1210** (1955).

3 E. R. van Driest, „On Turbulent Flow Near a Wall", *Journal of the Aeronautical Sciences*, November, 1007–1011, 1036 (1956).

4 E. R. Gilliland and T. K. Sherwood, „Diffusion of Vapors into Air Streams", *Industrial and Engineering Chemistry*, **26**, 516–523 (1934).

5 T. von Ka'rman, „The Analogy between Fluid Friction and Heat Transfer", *Transactions of the American Society of Mechanical Engineers*, **61**, 705–710 (1939).

6 W. M. Kays, „Heat Transfer to the Transpired Turbulent Boundary Layer", *International Journal of Heat and Mass Transfer*, **15**, 1023–1044 (1972).

7 H. S. Mickley, *NACA Technical Note*, **3208** (1954).

8 R. J. Moffat and W. M. Kays, „The Turbulent Boundary Layer on a Porous Plate: Experimental Heat Transfer with Uniform Blowing and Suction", *International Journal of Heat and Mass Transfer*, **11**, 1547–1566 (1968).

9 J. Nikuradse, VDI Forschungsheft, **356** (1932).

10 L. Prandtl, „Bericht über Untersuchungen zur ausgebildeten Turbulenz", *Zeitschrift für Angewandte Mathematik und Mechanik*, **5**, 136–139 (1925).

11 H. von Reichardt, „Vorträge aus dem Gebiet der Aero- und Hydrodynamik: Über das Messen turbulenter Längs- und Querschwankungen", *Zeitschrift für Angewandte Mathematik und Mechanik*, **18**, 358–361 (1938).

12 H. Schlichting, „*Boundary Layer Theory*", 6th edition, p. 560–595, 596–606, McGraw-Hill (1968).

9
Evaporation and Condensation

9.1
Characteristics of Simultaneous Heat and Mass Transfer

9.1.1
Mass Transfer with Phase Change

A gas at a temperature lower than its critical temperature is called a vapor and will easily condense to the liquid if the temperature becomes lower than the saturated temperature. Condensation, evaporation, humidification, and distillation are transport phenomena in which vapors or vapor mixtures are in contact with the liquids, and a large amount of energy is released or absorbed with the phase change, which is known as the *latent heat*. A characteristic of mass transfer in these phenomena is that it is always accompanied by an energy transfer due to the phase change, and in this respect it is quite different from ordinary mass transfer, which is not accompanied by an energy transfer. For this reason, we refer to *simultaneous heat and mass transfer*.

In this chapter, we discuss evaporation and condensation. Distillation, however, is discussed in the next chapter because of its practical importance, although the same principles as discussed herein can be applied to heat and mass transfer in distillation.

Figure 9.1 shows schematic representations of the temperature distributions near the interface between the gas and the liquid for the evaporation of pure liquids. Figure (a) shows the case with $T_L \gg T_G$, where the gas-phase sensible heat flux for heating the gas phase, q_G, and the latent heat flux for evaporation of the liquid, λN_A, are supplied from the liquid-phase sensible heat flux, q_L. Figure (b) shows the case with $T_L > T_G$, but where the gas-phase latent heat flux is supplied from the liquid-phase sensible heat flux, $q_L = \lambda N_A$. Figure (c) shows the case with $T_L \approx T_G$, where the latent heat flux for evaporation of the liquid, λN_A, is supplied from the sum of the liquid-phase and gas-phase sensible heat fluxes. The surface temperature of the liquid, T_s, is the lowest among $T_{G\infty}$, $T_{L\infty}$, and T_s, which is commonly the case for the evaporation of liquids. Figure (d) shows the case with $T_L < T_G$, where the latent heat flux, $\lambda_A N_A$, is supplied from the gas-phase sensible heat flux from the gas, q_G ($\lambda N_A = q_G$). The surface temperature in this special case

Mass Transfer. From Fundamentals to Modern Industrial Applications. Koichi Asano
Copyright © 2006 WILEY-VCH Verlag GmbH & Co. KGaA, Weinheim
ISBN: 3-527-31460-1

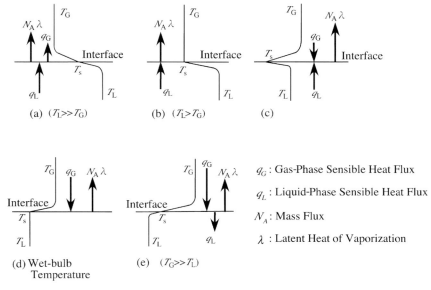

Fig. 9.1 Schematic representations of temperature distributions near the interface between the gas and the liquid for the evaporation of pure liquids.

is called the *wet-bulb temperature*. Figure (e) shows the case with $T_L \ll T_G$, where the gas-phase latent heat flux for evaporation, λN_A, and the liquid-phase sensible heat flux for heating the liquid, q_L, are supplied from the gas-phase sensible heat flux, q_G.

9.1.2
Surface Temperatures in Simultaneous Heat and Mass Transfer

If we define the direction from the liquid to the gas as being positive in describing the heat and mass fluxes, the energy balance at the vapor-liquid interface shown in Fig. 9.1 can be written by the following equation:

$$q_L = q_G + \lambda N_A \tag{9.1}$$

where N_A [kg m^{-2} s^{-1}] is the mass flux of the component A, q_G [W m^{-2}] is the gas-phase sensible heat flux, q_L [W m^{-2}] is the liquid-phase sensible heat flux, and λ [J kg^{-1}] is the latent heat of vaporization of component A.

From *the principle of local equilibrium*, the vapor pressures of the liquids at the vapor/liquid interface are in an equilibrium state and, according to the *phase rule*, they are dependent only on the interface temperature:

$$p_s = f(T_s) \tag{9.2}$$

Since the fluxes are proportional to their driving forces, we have the following relationships:

$$q_G \propto (T_s - T_{G\infty})$$

$$q_L \propto (T_s - T_{L\infty})$$

$$N_A \propto (p_s(T_s) - p_\infty)$$

Substituting the above expressions into Eq. (9.1), we obtain the following equation:

$$q_L(T_s) - q_G(T_s) - \lambda N_A(T_s) = 0 \tag{9.3}$$

Equation (9.3) indicates that the surface temperature can be estimated from the energy balance at the interface under the given bulk conditions, $(T_{L\infty}, T_{G\infty}, p_{G\infty})$. Once the surface temperature is known, the sensible heat fluxes and mass fluxes can be estimated by the same methods as described in the previous chapters. Since heat and mass transfer are interrelated through the boundary conditions, Eq. (9.1) is characteristic of *simultaneous heat and mass transfer*.

9.2
Wet-Bulb Temperatures and Psychrometric Ratios

In the following, we discuss a special case of simultaneous heat and mass transfer, namely the issue of wet-bulb temperatures. From the energy balance shown in Fig. 9.1 (d), we have the following equation:

$$q_G = \lambda_w N_A \tag{9.4}$$

where N_A [kg m^{-2} s^{-1}] is the gas-phase mass flux, q_G [W m^{-2}] is the gas-phase sensible heat flux, and λ_w [J kg^{-1}] is the latent heat of vaporization of the liquid at the wet-bulb temperature T_w [K].

The gas-phase sensible heat flux and the mass flux can be expressed by the following equations:

$$q_G = h_G(T_{G\infty} - T_w) \tag{9.5}$$

$$N_A = k_H(H_w - H_\infty) \tag{9.6}$$

where h_G [W m^{-2} K^{-1}] is the gas-phase heat transfer coefficient, k_H [kg m^{-2} s^{-1}] is the gas-phase mass transfer coefficient defined by the driving force in terms of the absolute humidity, H_∞ is the absolute humidity of the bulk gas, H_w is the absolute humidity at the wet-bulb temperature, and T_w and $T_{G\infty}$ are the wet-bulb and the bulk gas temperatures [K], respectively.

Substituting Eqs. (9.5) and (9.6) into Eq. (9.4) and rearranging, we obtain the following equation:

$$\frac{H_w - H_\infty}{T_{G\infty} - T_w} = \frac{1}{\lambda_w}\left(\frac{h_G}{k_H}\right) \tag{9.7}$$

The parameter (h_G/k_H) on the right-hand side of Eq. (9.7) is known as the *psychrometric ratio* [kJ kg^{-1} K^{-1}], and is one of the physical properties of the system that will be discussed below. Equation (9.7) is the theoretical foundation of dry- and wet-bulb hygrometers. For the air/water system, the recommended value of the psychrometric ratio is 1.09 kJ kg^{-1} K^{-1} (= 0.26 kcal kg^{-1} K^{-1}).

Let us consider the physical meaning of the psychrometric ratio. If we assume that the laminar boundary layer theory along a flat plate is applicable for heat and mass transfer in the evaporation of liquids, we have the following equation:

$$Nu_G = \frac{h_G L}{\kappa_G} = 0.664 Pr^{1/3} Re^{1/2} \tag{4.40}$$

$$Sh(1-\omega_w) = \frac{k_H L(1-\omega_w)}{\rho_G D_G}\left(\frac{H_w - H_\infty}{\omega_w - \omega_\infty}\right) = 0.664 Sc^{1/3} Re^{1/2} \tag{4.41}$$

From Tab. 2.1.b, we have the following equation:

$$H = \frac{\omega}{1-\omega} \tag{9.8}$$

Substituting Eqs. (4.40), (4.41), and (9.8) into Eq. (9.7), we have the following equation:

$$\frac{h_G}{k_H} = c_p Le^{2/3}(1-\omega_w)\left(\frac{H_w - H_\infty}{\omega_w - \omega_\infty}\right) = c_p Le^{2/3}\frac{1}{1-\omega_\infty} \tag{9.9}$$

where Le is the *Lewis number* defined by the following equation:

$$Le \equiv \frac{Sc}{Pr} = \left(\frac{\kappa_G}{\rho_G c_p D_G}\right) \tag{9.10}$$

Here, c_p is the specific heat at constant pressure [J kg^{-1} K^{-1}], D_G is the diffusivity [m^2 s^{-1}], κ_G is the thermal conductivity [W m^{-1} K^{-1}], and ρ_G is the density [kg m^{-3}].

If the vapor concentration in the gas phase is very low, $\omega_\infty \approx 0$, as is the case for humid air, we can approximate Eq. (9.9) by the following equation:

$$\frac{h_G}{k_H} \approx c_p Le^{2/3} \tag{9.11}$$

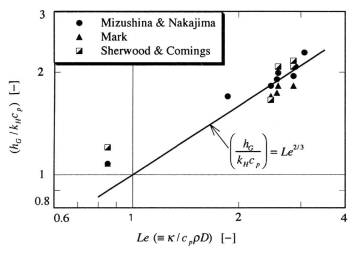

Fig. 9.2 Comparison of the observed psychrometric ratio data obtained by previous workers with Eq. (9.11); data from T. Mizushina and M. Nakajima [8], J. G. Mark [11], and T. K. Sherwood and E. W. Comings [11].

Equation (9.11) indicates that the psychrometric ratio is a physical property of the system.

Figure 9.2 shows a comparison of the observed psychrometric ratio data obtained by previous workers [8, 11] with Eq. (9.11). The solid line in the figure represents Eq. (9.11), and can be seen to show fairly good agreement with the data.

Example 9.1

The dry- and wet-bulb temperatures of humid air at 1 atm (101.325 kPa) are 300 K and 293 K, respectively. Calculate the absolute humidity of the air by use of the psychrometric ratio of humid air, h_G/k_H, of 1.09 kJ kg^{-1} K^{-1}.

Solution

From the steam table, we have the following values:

$$\lambda_{293K} = 2454.7 \text{ kJ kg}^{-1}, \; p_{293K} = 2.32 \text{ kPa}, \; p_{300K} = 3.53 \text{ kPa}$$

The absolute humidity at the wet-bulb temperature:

$$H_w = \frac{2.32}{101.325 - 2.32} \left(\frac{18.02}{28.97}\right) = 0.0146$$

Substituting these values into Eq. (9.7), we have:

$$\frac{0.0146 - H_\infty}{300 - 293} = \frac{1.09}{2454.7}$$

$$H_\infty = 0.0146 - (1.09)(300 - 293)/(2454.7) = 0.0115 \text{ kg/kg air}$$

$$\omega_\infty = \frac{0.0115}{1 + 0.0115} = 0.0114$$

The partial pressure of the water vapor in the humid air p [kPa] and the relative humidity are given by:

$$p = \frac{(0.0114/18.02)(101.325)}{(0.9886/28.97) + (0.0114/18.02)} = 1.84 \text{ [kPa]}$$

$$\text{relative humidity} = (100)(1.84)/(3.53) = 52.1\%$$

Example 9.2
A gasoline drop of diameter 0.1 mm is falling through air at 323 K and 1 atm at a velocity of 0.15 m s^{-1}. Calculate the surface temperature of the drop, if the physical properties of gasoline are approximated by those of hexane.

Solution
The recommended approach to this problem is to apply the simulation technique as described in Example 7.1, but here we will apply a rather approximate method based on the use of the psychrometric ratio according to Eq. (9.7).

The physical properties of the air/hexane system are as follows:

$$Le = 1.85$$

Specific heat at the drop surface:

$$c_p = 1.12 \text{ kJ kg}^{-1} \text{ K}^{-1}$$

Estimation of the psychrometric ratio by use of Eq. (9.11):

$$h_G/k_H = (1.12)(1.85)^{2/3} = 1.69 \text{ kJ kg}^{-1} \text{ K}^{-1}$$

Saturated vapor pressure of hexane:

$$\log p = 5.99514 - \frac{1168.72}{T - 48.94} \qquad (p \text{ [kPa]}, \ T \text{ [K]}) \qquad \text{(A)}$$

$$H_w = \frac{(p/101.3)}{1 - (p/101.3)}\left(\frac{86.18}{28.97}\right) \qquad \text{(B)}$$

Latent heat of vaporization of hexane:

$$\lambda = 366 + 0.716\,(T - 298.15) \text{ kJ kg}^{-1} \tag{C}$$

Substituting the above values into Eq. (9.7), we have:

$$T_{w,cal} = 323 - \frac{H_w \lambda_w}{1.69} \tag{D}$$

The wet-bulb temperature is calculated by the use of Eqs. (A)–(D) on a trial-and-error basis. The calculation starts by tentatively assuming the initial values of T_w in Eq. (A), which gives the first approximation, $T_{w,cal}$. The calculation is repeated until convergence in the value of $T_{w,cal}$ is obtained. The final result is $T_w = 275.1$ K, as shown in Tab. 9.1. This value is lower than that obtained in Example 7.1 ($T_w = 277.5$ K) by 2.4 K. This may be due to the fact that Eq. (9.11) is obtained by a laminar boundary layer approximation, whereas in Example 7.1 it is based on a rigorous numerical simulation.

Table 9.1 Calculation of wet-bulb temperature in Example 9.2.

T_w [K]	p [kPa]	H_w	λ [kJ kg^{-1}]	T_{wcal} [K]
300.0	21.878	0.8192	364.7	146.2
290.0	14.024	0.4779	371.8	217.9
280.0	8.651	0.2777	379.0	260.7
270.0	5.108	0.1579	386.2	286.9
275.0	6.686	0.2102	382.6	275.4
275.2	6.757	0.2126	382.4	274.9
275.1	6.722	0.2114	382.5	275.1

9.3
Film Condensation of Pure Vapors

9.3.1
Nusselt's Model for Film Condensation of Pure Vapors

There are two types of condensation of vapors, namely film condensation, in which the condensate wets the surface, spreads out, and establishes a stable film over the condensing surface, and dropwise condensation, in which the condensate does not wet the condensing surface but stays adhered as separate drops. In this section, we confine our discussions to the film condensation of pure vapors, which is a fundamental concern in the design of industrial condensers.

W. von Nusselt [9] proposed his famous model for the condensation of pure vapors on a vertical flat plate on the basis of the following assumptions:
1. The vapor phase consists of stationary saturated vapor.
2. The effect of vapor shear on the surface of the condensate is negligibly small.
3. The temperature distribution in the liquid film is linear.
4. The physical properties are constant.

Figure 9.3 shows a physical representation of the film condensation of a pure vapor on a vertical flat plate. The velocity in the liquid film flowing along a vertical flat plate is very small and we can neglect the inertia term in the equation of motion. We then obtain the following equation:

$$\mu_L \frac{d^2 v_z}{dy^2} + g(\rho_L - \rho_G) = 0 \tag{9.12}$$

where μ_L [Pa s] is the viscosity of the liquid, v_z [m s^{-1}] is the velocity of the liquid at z [m] from the leading edge of the condensing wall, ρ_G and ρ_L [kg m^{-3}] are the density of the vapor and the liquid, respectively, δ [m] is the thickness of the liquid film, and y [m] is the distance from the condensing wall.

The boundary conditions for the above equation can be written as follows:

$y = 0: \quad v_z = 0$

$y = \delta: \quad dv_z/dy = 0$

Integrating Eq. (9.12) with respect to y, we obtain the following equation:

$$v_z = \frac{g(\rho_L - \rho_G)}{\mu_L}\left(\delta y - \frac{y^2}{2}\right) \tag{9.13}$$

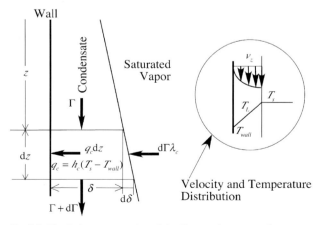

Fig. 9.3 Physical representation of the film condensation of a pure vapor on a vertical flat plate.

9.3 Film Condensation of Pure Vapors

The flow rate of the liquid per unit width of the condensing wall, Γ [kg m^{-1} s^{-1}], is given by the following equation:

$$\Gamma \equiv \rho_L \int_0^\delta v_z dy = \frac{g\rho_L(\rho_L - \rho_G)}{\mu_L} \frac{\delta^3}{3} \tag{9.14}$$

In film condensation, the vapor is condensed at the interface and is cooled to the average temperature of the liquid. The change in specific enthalpy with condensation, λ_c [J kg^{-1}], can be expressed as the sum of the enthalpy change at the vapor-liquid interface (heat of vaporization at the interface temperature) and that due to the difference between the interface temperature and the average temperature of the condensate. If we assume a linear temperature distribution in the liquid film, the change in enthalpy with condensation can be expressed by the following equation:

$$\lambda_c = \lambda_s + \frac{1}{\Gamma} \int_0^\delta \rho_L v_z c_L (T_s - T_{wall}) dy = \lambda_s + \frac{3}{8} c_L (T_s - T_{wall}) \tag{9.15}$$

where c_L is the specific heat of the liquid [J kg^{-1} K^{-1}], T_s is the interface temperature [K], T_{wall} is the wall temperature (the surface temperature of the condensing wall) [K], and λ_s is the latent heat of vaporization [J kg^{-1}] of the vapor at the interface temperature T_s.

If we assume that the transfer of the sensible heat in the liquid film in the z-direction is negligibly small in comparison with that in the y-direction, we have the following equation from the energy balance for a small element of the liquid, dz:

$$q_{cz} = \kappa_L \frac{(T_s - T_{wall})}{\delta} = \lambda_c \frac{d\Gamma}{dz} \tag{9.16}$$

Substituting Eq. (9.14) into Eq. (9.16) and rearranging, we have the following equation:

$$\delta^3 d\delta = \frac{\kappa_L \mu_L (T_s - T_{wall})}{g\rho_L(\rho_L - \rho_G)\lambda_c} dz \tag{9.17}$$

Integrating Eq. (9.17) with respect to z under the following boundary conditions:

$$z = 0: \quad \delta = 0$$

we obtain the following equation for the thickness of the liquid film, δ:

$$\delta = \left\{ \frac{4\kappa_L \mu_L (T_s - T_{wall}) z}{g\rho_L(\rho_L - \rho_G)\lambda_c} \right\}^{1/4} \tag{9.18}$$

From Eq. (9.18), we have the following equation for the local *condensation heat transfer coefficient* h_c [W m^{-2} K^{-1}]:

$$h_{cz} \equiv \frac{\kappa_L}{\delta} = \left\{ \frac{g\rho_L(\rho_L - \rho_G)\kappa_L^3 \lambda_c}{4\mu_L(T_s - T_{wall})z} \right\}^{1/4} \tag{9.19}$$

The average condensation heat transfer coefficient over the total length of the condensing wall L [m], $\overline{h_c}$ [W m^{-2} K^{-1}], is given by the following equation:

$$\overline{h_c} \equiv \frac{1}{L}\int_0^L h_{cz}\,dz = 0.943\left\{ \frac{g\rho_L(\rho_L - \rho_G)\kappa_L^3 \lambda_c}{\mu_L(T_s - T_{wall})L} \right\}^{1/4} \tag{9.20}$$

Equations (9.19) and (9.20) are known as *Nusselt's model* for the film condensation of pure vapors.

The average heat flux over the total length of the condensing wall, $\overline{q_L}$ [W m^{-2}], is given by the following equation:

$$\overline{q_L} = \overline{h_c}(T_s - T_{wall}) \tag{9.21}$$

The rate of condensation of the vapor per unit width of the condensing wall, Γ [kg m^{-2} s^{-1}], is given by:

$$\Gamma = \overline{q_L} L/\lambda_s \tag{9.22}$$

Nusselt's model can be used to predict the rates of condensation of pure vapors fairly accurately, if the physical properties of the liquid and the temperature at the vapor-liquid interface are evaluated in a suitable way, in which case agreement between experimental data and the model is very good.

An important fact about Nusselt's model is that it is based on the energy balance for the liquid film and does not rely in any way on the definition of the interface temperature. In the case of the condensation of a saturated pure vapor, the interface temperature, T_s, is always equal to the bulk vapor temperature, $T_{G\infty}$, as is usually assumed. In the case of the condensation of a vapor in the presence of a non-condensable gas, however, the latter accumulates near the interface, as will be discussed in Section 9.5, and this causes an additional thermal resistance in the vapor phase and the assumption of $T_{G\infty} = T_s$ no longer holds. In the case of the condensation of vapor mixtures, the temperature difference due to the vapor-liquid equilibrium causes another type of thermal resistance in the vapor phase. Nevertheless, even in such cases, Nusselt's model can be used to predict the liquid-phase sensible heat fluxes very well, if the true interface temperature is estimated. The problem of condensation in the presence of a non-condensable gas or the condensation of vapor mixtures thus reduces to a problem of estimating the interface temperature.

9.3.2
Effect of Variable Physical Properties

The Nusselt model is derived under the assumption of constant physical properties of the liquid. In reality, however, the physical properties of a liquid are very sensitive to variations in temperature, especially its viscosity. Furthermore, there usually exists a large temperature difference between the vapor-liquid interface and the surface of the condensing wall. This presents us with a serious problem in applying Nusselt's model.

Table 9.2 Temperature dependence of the viscosity and thermal conductivity of water at 373.15 K and 298.15 K.

Location	Temperature [K]	Viscosity [mPa s]	Thermal conductivity [W m^{-1} K^{-1}]
heat transfer surface	298.15	0.8902	0.609
interface	373.15	0.2822	0.681

Let us consider, for example, the case of condensation of saturated steam at 1 atm (T_s = 373.15 K) on a condensing wall at 298.15 K. Table 9.2 summarizes the physical properties of the condensate at the interface and at the condensing wall. The condensation heat transfer coefficient, h_c, calculated from the physical properties at the condensing wall, T_{wall}, shows a 40% higher value than that calculated from the physical properties at the interface temperature, T_s.

On the basis of their detailed and elaborate numerical calculations, W. J. Minkowycz and E. M. Sparrow [7] recommend the use of the *reference temperature* for this problem, which is defined by the following equation:

$$T_{ref} = T_{wall} + 0.31\,(T_s - T_{wall}) \tag{9.23}$$

They reported that the Nusselt model calculated from the physical properties at the reference temperature by means of Eq. (9.23) showed good agreement with the results of a numerical calculation taking into account the effect of variable physical properties.

Example 9.3
Saturated steam at 1 atm ($T_s = T_{G\infty}$ = 373.15 K) condenses on the surface of a vertical flat plate of length 1 m and width 0.1 m. Calculate the rate of condensation of the steam [kg s^{-1}] if the surface temperature of the wall is kept at 298.15 K.

If the condensing wall is replaced with one of length 0.1 m and width 1 m, how will the rate of condensation be affected?

Solution
The physical properties under the specified conditions are as follows:

The reference temperature is given by Eq. (9.23):

$T_{ref} = 298.15 + 0.31(373.15 - 298.15) = 321.4$ K

The liquid properties at 321.4 K are as follows:

$\rho_L = 988.9$ kg m^{-3}, $\mu_L = 5.64 \times 10^{-4}$ Pa s, $\kappa_L = 0.639$ W m^{-1} K^{-1}, $c_L = 4.18$ kJ kg^{-1} K^{-1}

The vapor properties at 373.15 K are as follows:

$\rho_G = 0.684$ kg m^{-3}, $\lambda_s = 2256.9$ kJ kg^{-1}

Substituting the above values into Eq. (9.15), we have:

$\lambda_c = 2256.9 + (3/8)(4.18)(373.15 - 298.15) = 2374.5$ kJ kg^{-1}

Substituting these values into Eq. (9.20), we have:

$$\overline{h_c} = (0.943)\left\{\frac{(9.81)(988.9)(988.9 - 0.68)(0.639)^3(2.3745 \times 10^6)}{(5.64 \times 10^{-4})(373.15 - 298.15)(1.00)}\right\}^{1/4}$$

$= 3.24$ kWm^{-2} s^{-1}

Rate of condensation heat transfer:

$Q_c = (3.24)(1.0)(0.1)(373.15 - 298.15) = 24.3$ kW

Rate of condensation of vapor:

$W = (24.3)/(2374.5) = 0.0102$ kg s^{-1}

In the latter case, the heat transfer surface is the same, but the total length of the condensing wall is only 1/10th of that in the former case. From Eq. (9.20), the ratio of the condensation heat transfer coefficients is given by:

$(h_{c2}/h_{c1}) = (10)^{1/4} = 1.78$

The rate of condensation is thus 1.78 times larger than in the former case:

$W_2 = (1.78)(0.0102) = 0.0182$ kg s^{-1}

It is for this reason that the horizontal-type total condenser is used in actual plants.

9.4
Condensation of Binary Vapor Mixtures

9.4.1
Total and Partial Condensation

The condensation of vapor mixtures is a fundamental concern for many industrially important pieces of apparatus, such as the total condensers of distillation columns. In this section, we discuss some characteristics of heat and mass transfer in the condensation of binary vapor mixtures.

Figure 9.4 shows a schematic representation of the condensation of a binary vapor mixture at constant total pressure. The ordinates are the vapor and liquid temperatures. The abscissae are the vapor and liquid concentrations in terms of mole fraction. The curve AFE represents the dew-point curve and the curve DCB the bubble-point curve. At the interface between the vapor and the liquid, the vapor and liquid are in equilibrium with each other according to the *principle of local equilibrium*. T_D is the dew point of the vapor of concentration x_c and T_B is the bubble point of the liquid of concentration x_c. If the saturated vapor at point A (x_c, T_D) comes into contact with the cold surface at a temperature sufficiently lower than the bubble point, T_B, the concentration of the condensed liquid will be the same as that of the bulk vapor. The point B represents the state of the liquid at the interface and the point E that of the equilibrium vapor at the interface. This type of condensation is called *total condensation*.

If, on the other hand, the temperature of the condensing surface is not cold enough, the interface temperature, T_s, will be higher than the bubble point of the liquid, $T_B(x_c)$, yet lower than the dew point of the vapor, T_D, ($T_D(x_c) > T_s > T_B(x_c)$), and the concentration of the condensed liquid will always be lower than that of the bulk vapor ($x_c > x_s$). This type of condensation is called *partial condensation*. The point C represents the state of the liquid at the interface and the point F that

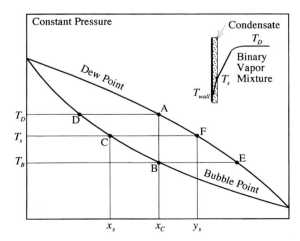

Fig. 9.4 Physical representation of the total and partial condensation of binary vapor mixtures.

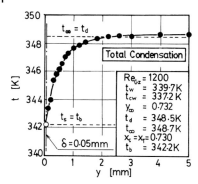

Fig. 9.5 Vapor-phase temperature distribution in the total condensation of binary vapor mixtures; data from K. Asano and S. Matsuda [2].

of the equilibrium vapor. If the temperature of the condensing wall becomes higher than T_D, no condensation of vapor will take place. To sum up, a characteristic feature of the condensation of binary vapor mixtures is that there are two types of condensation, namely total condensation and partial condensation, according to the cooling conditions.

9.4.2
Characteristics of the Total Condensation of Binary Vapor Mixtures

K. Asano and S. Matsuda [2] presented an experimental approach to the total and partial condensation of binary vapor mixtures on a vertical flat plate under forced-flow conditions using the methanol/water system as an example. Figure 9.5 shows an example of an observed vapor-phase temperature distribution for the total condensation of a binary vapor consisting of 73.2% methanol (T_{dew} = 348.5 K) in contact with a wall at 339.7 K at a total pressure of 1 atm. The solid line in the figure represents the theoretical values derived from laminar boundary layer theory with surface mass suction, and can be seen to show good agreement with the observed data. This suggests that the vapor-phase heat and mass fluxes in the condensation of binary vapor mixtures on a vertical flat plate under forced-flow conditions can be estimated by laminar boundary layer theory with surface mass suction as described in Section 4.8. The observed bulk vapor temperature (348.7 K) is nearly equal to the dew point of the vapor (348.5 K) and the observed interface temperature agrees well with the bubble point of the liquid under total condensation (342.2 K). The relative orders of magnitude of the thermal resistances for the case shown in Fig. 9.5 are as follows:

Total resistance: $T_{G\infty} - T_{wall} = 348.7 - 339.7 = 9.0$ K (100%)

Vapor-phase resistance: $T_{G\infty} - T_s = 348.7 - 342.2 = 6.5$ K (72%)

Liquid-phase resistance: $T_s - T_{wall} = 342.2 - 339.7 = 2.5$ K (28%)

This points to an important fact about the total condensation of a binary vapor, i.e. that there is a large temperature difference in the vapor phase and hence we

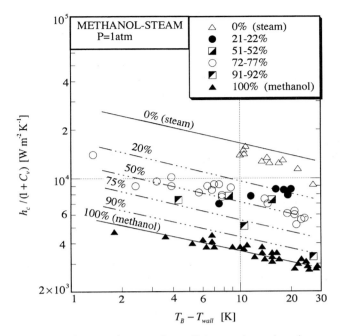

Fig. 9.6 Condensation heat transfer coefficients under total condensation of binary vapor mixtures [2].

cannot use the bulk vapor temperature, $T_{G\infty}$, as the interface temperature of the liquid. Instead, we must use the bubble point of the condensate, $T_B(x_c)$, as the interface temperature:

$$T_s = T_B(x_c)$$

9.4.3
Rate of Condensation of Binary Vapors under Total Condensation

Figure 9.6 shows the observed condensation heat transfer coefficients, h_c [W m^{-2} K^{-1}], for the total condensation of methanol/water systems of various concentrations on a short vertical flat plate [2]. The ordinate is the condensation heat transfer coefficient defined by the temperature difference between the interface and the wall, $T_s - T_{wall}$, corrected for the effect of vapor shear, and the abscissa is the temperature difference in the liquid film ($T_s - T_{wall}$), where T_s is taken to be equal to the bubble point of the condensate $T_B(x_c)$. The two solid lines in the figure represent the theoretical values based on the Nusselt model for the condensation of pure methanol and water vapor, respectively. The four dashed lines are the theoretical values obtained by the Nusselt model for the total condensation of mixtures containing 20%, 50%, 75%, and 90% methanol vapor, where the interface temperatures are taken to be equal to the bubble points of the liquids, $T_s = T_B(x_c)$, and the physical properties of the liquids are estimated by the use of Kay's method at the reference temperatures based on Eq. (9.23). The fact that good agreement is observed between experimental data and theory indicates that the Nusselt model is also applicable for the total condensation of mixed vapors, if the bubble point of the liquid is taken as the surface temperature of the condensate.

9.5
Condensation of Vapors in the Presence of a Non-Condensable Gas

9.5.1
Accumulation of a Non-Condensable Gas near the Interface

If a vapor is contaminated with a small amount of a non-condensable gas, the transport mechanism of the condensation becomes vapor-phase mass transfer controlled rather than liquid-phase heat transfer controlled. Figure 9.7 shows a physical representation of the film condensation of a vapor in the presence of a non-condensable gas.

Here, we consider the case of condensation at constant total pressure P under forced-flow conditions. If the temperature of the bulk vapor is $T_{G\infty}$ and the interface temperature of the liquid is T_s ($T_{G\infty} > T_s$), the vapor pressure in the bulk vapor is $p_v(T_{G\infty})$ and that at the interface is $p_v(T_s)$. We then have the following equations:

Bulk vapor: $\quad P = p_v(T_{G\infty}) + p_{G\infty}$ \hfill (9.24a)

Interface: $\quad P = p_v(T_s) + p_{Gs}$ \hfill (9.24b)

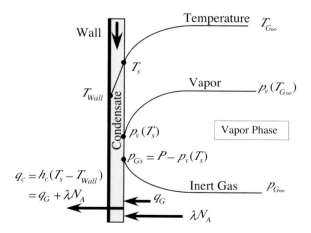

Fig. 9.7 Physical representation of the condensation of a vapor in the presence of a non-condensable gas.

From Eqs. (9.24a) and (9.24b) we have the following equation:

$$p_{Gs} - p_{G\infty} = p_v(T_{G\infty}) - p_v(T_s) \geq 0 \qquad (9.25)$$

If the vapor condenses at the interface, the mass flux due to condensation must be compensated by the diffusional flux of the vapor from the bulk vapor to the interface, which means that $p_v(T_{G\infty})$ should be larger than $p_v(T_s)$ and that $T_{G\infty}$ is higher than T_s. Equation (9.25) suggests that the non-condensable gas will accumulate near the interface, thereby creating a further thermal resistance in the vapor phase.

9.5.2
Calculation of Heat and Mass Transfer

Figure 9.7 suggests that laminar boundary layer theory with surface mass suction, as described in Section 4.8, may be applied to vapor-phase heat and mass transfer in the film condensation of vapors in the presence of a non-condensable gas under forced flow. Since the relative order of magnitude of the vapor-phase sensible heat flux, q_G, is very small in comparison with that of the latent heat flux, λN_A, we further assume that $B_H \approx B_M$, which greatly simplifies the numerical calculation, and we obtain the following equations for the vapor-phase heat and diffusional flux:

Vapor-phase local heat flux:

$$Nu_z = 0.332\, Re_z^{1/2}\, Pr^{1/3}\, g(B_M) \qquad (9.26)$$

Vapor-phase local diffusional flux:

$$Sh_z(1 - \omega_s) = 0.332\, Re_z^{1/2}\, Sc^{1/3}\, g(B_M) \qquad (4.104)$$

$$g(B_M) = \frac{1}{0.09 + 0.91(1 + B_M)^{0.8}} \qquad (4.105)$$

where $Re_z (\equiv \rho_G U_\infty z / \mu_G)$ is the local Reynolds number and B_M is the transfer number for mass transfer:

$$B_M = \frac{\omega_s - \omega_\infty}{1 - \omega_s} \qquad (4.108)$$

For the liquid-phase local heat transfer, we can apply Nusselt's model using Eq. (9.19).

The interface temperature, which is necessary for prediction of the rate of mass transfer, can also be obtained from the energy balance at the interface, as is the case for the evaporation of liquids:

$$q_L(z) = q_G(z) + \lambda N_A(z) \qquad (9.27)$$

9.5.3
Experimental Approach to the Effect of a Non-Condensable Gas

D. F. Othmer [10] was the first to present an experimental approach to describing the effect of a non-condensable gas on the rate of condensation of a vapor. He studied the condensation of stationary steam mixed with small amounts of air on a horizontal tube at constant pressure. Figure 9.8 shows the results of Othmer's experiment, which indicate that contamination with even a few percent of air will cause a considerable decrease in the rate of condensation of steam. The apparent decrease in the condensation heat transfer coefficient, $\overline{h_c}$, may be due to the effect of accumulation of air near the interface, which causes large temperature differences in the vapor phase ($T_{G\infty} > T_s$). Othmer, however, assumed that $T_s = T_{G\infty}$ in his data processing, which inevitably led to overestimation of the temperature driving force in the liquid film.

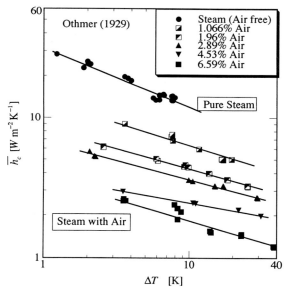

Fig. 9.8 Othmer's experiment [10] on the effect of a non-condensable gas on condensation heat transfer (under natural convection).

K. Asano et al. [1] presented an experimental approach to heat and mass transfer in the condensation of various vapors, such as steam, benzene, methanol, and carbon tetrachloride, on a short vertical flat plate under forced-flow conditions and at constant total pressure. Figure 9.9 shows a comparison of the observed data with theoretical values obtained by laminar boundary layer theory with surface mass suction. In the figure, the ordinate represents the dimensionless vapor-phase average diffusional flux $\overline{Sh}(1 - \omega_s)$, and the abscissa is the transfer number

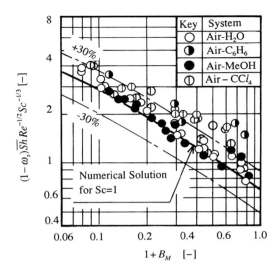

Fig. 9.9 Effect of a non-condensable gas on condensation on a vertical flat plate under forced-flow conditions; data from K. Asano et al. [1].

for mass transfer B_M. The solid line in the figure represents the vapor-phase average diffusional flux obtained from laminar boundary layer theory with surface mass suction:

$$\overline{Sh}(1 - \omega_s) = 0.664 \, Re^{1/2} \, Sc^{1/3} \, g(B_M) \qquad (4.107)$$

where $g(B_M)$ is given by Eq. (4.105)

The dimensionless diffusional fluxes evaluated from the observed interface temperatures, $T_{s,ob}$, showed good agreement with the theoretical values derived from Eq. (4.107). The observed liquid-phase heat transfer coefficients evaluated from the temperature difference between the interface and the wall, $(T_{s,ob} - T_{wall})$, also showed good agreement with the theoretical values obtained from Nusselt's model. These facts suggest that rates of condensation in the presence of a non-condensable gas can easily be predicted if the interface temperatures are suitably evaluated.

Example 9.4

Steam mixed with 5% air at 1 atm and 371.72 K is supplied to a vertical flat plate condenser. The condensing wall is 0.2 m in length and 0.05 m in width and the surface temperature of the wall is kept at 298.15 K. Calculate the rate of condensation if the vapor is supplied at a downward velocity of 0.5 m s^{-1}. Assume sufficient supply of the vapor such that the velocity remains essentially the same even at the end of the condenser.

Solution

Step 1: Estimation of the physical properties:

The physical properties of the system are estimated by application of the following equations:

Vapor pressure of the steam:

$$\log p = 10.09170 - 1668.21/(T - 45.15)$$

Viscosity of air and steam (Hirschfelder's equation):

$$\mu = \frac{2.669 \times 10^{-6} \sqrt{MT}}{\sigma^2 \Omega_V (Tk/\varepsilon)} \quad [\text{Pa} \cdot \text{s}]$$

Thermal conductivity of air and steam (modified Eucken's equation):

$$\kappa = (18.68 + C_p/\gamma)(1000 \mu/M) \quad [\text{W m}^{-1} \text{ K}^{-1}]$$

Viscosity and thermal conductivity of the vapor mixture (Wilke's equation):

$$\mu_{Gmix} = \frac{\mu_1 y_1}{y_1 + y_2 \phi_{12}} + \frac{\mu_2 y_2}{y_2 + y_1 \phi_{21}} \quad [\text{Pa} \cdot \text{s}]$$

$$\kappa_{Gmix} = \frac{\kappa_1 y_1}{y_1 + y_2 \phi_{12}} + \frac{\kappa_2 y_2}{y_2 + y_1 \phi_{21}} \quad [\text{W m}^{-1} \text{ K}^{-1}]$$

Diffusivity of steam (Hirschfelder's equation):

$$D = \frac{1.858 \times 10^{-7} T^{3/2} \sqrt{1/M_A + 1/M_B}}{(P/101325) \sigma_{AB}^2 \Omega_D (kT/\varepsilon_{AB})} \quad [\text{m}^2 \text{ s}^{-1}]$$

Table 9.3 summarizes the parameters used to estimate the physical properties of the vapor.

Table 9.3 Parameters for estimation of the physical properties of vapors.

Parameters	Unit	Air	Steam
molecular weight	[kg kmol^{-1}]	28.97	18.02
σ	[Å]	3.62	2.65
ε/k	[K]	92	356
molar heat a	[J mol^{-1} K^{-1}]	28.06	32.24
b	[J mol^{-1} K^{-2}]	0.00197	0.00192
c	[J mol^{-1} K^{-3}]	4.80E-06	1.06E-05
d	[J mol^{-1} K^{-4}]	−1.90E-09	−3.61E-09
C_p/C_v	[−]	1.400	1.324

9.5 Condensation of Vapors in the Presence of a Non-Condensable Gas

Physical properties of liquid:

Latent heat of vaporization of water:

$\lambda = 3150 - 2.37\, T$ [kJ kg^{-1}]

Viscosity of water:

$\mu_L = \exp(-3.254 + 429.81/(T - 161.3))$ [mPa s]

Thermal conductivity of water:

$\kappa_L = -0.258 + 4.40 \times 10^{-3}\, T - 5.00 \times 10^{-6}\, T^2$ [W m^{-1} K^{-1}]

Specific heat of water:

$c_L = 4.18$ kJ kg^{-1} K^{-1}

Step 2: Calculation of heat and mass fluxes:

Vapor-phase heat and diffusional fluxes are calculated by applying laminar boundary layer theory with surface mass suction. The liquid-phase sensible heat flux is calculated by means of the Nusselt model.

Vapor-phase sensible heat flux:

$Nu_z = 0.332\, Pr^{1/3}\, Re_z^{1/2}\, g(B_M)$

$q_{Gz} = Nu_z \{\kappa_{Gs}(T_{G\infty} - T_s)/z\}$

Vapor-phase mass flux:

$Sh_z(1 - \omega_s) = 0.332\, Sc^{1/3}\, Re_z^{1/2}\, g(B_M)$

$N_{Az} = Sh_z B_M (\rho_{Gs} D_{Gs}/z)$

High mass flux is estimated by means of Eq. (4.105):

$g(B_M) = 1/\{0.09 + 0.91(1 + B_M)^{0.8}\}$

Liquid-phase sensible heat flux is given by Eq. (9.19):

$h_{cz} = \left\{ \dfrac{g\rho_L(\rho_L - \rho_G)\kappa_L^3 \lambda_c}{4\mu_L(T_s - T_{wall})z} \right\}^{1/4}$

$q_{Lz} = h_{cz}(T_s - T_{wall})$

Step 3: Estimation of the interface temperature T_s:

The interface temperature at each position on the condensing wall is evaluated so as to satisfy the following boundary condition at the interface:

$q_{Lz} - q_{Gz} - \lambda_z N_{Az} = 0$

Step 4: Calculation of the flow rate of the condensate and the heat transfer rate:

$$\Gamma \equiv \int_0^z N_A dz = \sum (z_i - z_{i-1})(N_{A,i} + N_{A,i-1})/2$$

$$Q \equiv \int_0^z q_L dz = \sum (q_{L,i} + q_{L,i-1})(z_i - z_{i-1})/2$$

Table 9.4 summarizes the results of numerical calculations based on the above equations. From Tab. 9.4, the flow rate of the condensed liquid at the lower end of the condensing wall and the total rate of heat transfer are given by:

$\Gamma_{bottom} = 3.88 \times 10^{-3}$ kg m^{-1} s^{-1}

$Q_{bottom} = 10.0$ kW m^{-1}

Table 9.4a Condensation of steam with 5% air on a vertical flat plate at a total pressure of 1 atm in Example 9.4.

z [m]	T_s [K]	B [−]	Pr [−]	Sc [−]	Re_z [−]	Nu_z [−]	$Sh_z(1-\omega_s)$ [−]
0.00	371.72						
0.005	310.63	−0.919	0.759	0.612	78.0	12.6	11.7
0.01	308.16	−0.919	0.748	0.613	156	17.8	16.6
0.02	306.16	−0.919	0.756	0.614	312	25.3	23.6
0.04	304.54	−0.920	0.753	0.614	624	35.7	33.4
0.06	303.75	−0.920	0.752	0.615	935	43.8	40.9
0.08	303.25	−0.920	0.751	0.615	1247	50.6	47.3
0.10	302.89	−0.920	0.751	0.615	1559	56.5	52.9
0.12	302.61	−0.920	0.750	0.615	1871	61.9	58.0
0.14	302.39	−0.920	0.750	0.615	2182	66.9	62.6
0.16	302.21	−0.920	0.750	0.615	2494	71.5	67.0
0.18	302.05	−0.920	0.750	0.615	2806	75.9	71.0
0.20	301.92	−0.920	0.750	0.615	3118	80.0	74.9

9.5 Condensation of Vapors in the Presence of a Non-Condensable Gas

Table 9.4b Example 9.4 continued.

z [m]	q_L [kW m^{-2}]	q_G [kW m^{-2}]	$N_A\lambda_s$ [kW m^{-2}]	N_A [kg m^{-2} s^{-1}]	Γ [kg m^{-1} s^{-1}]	$Q/Q_{Nusselt}$ [–]	$\dfrac{T_s - T_{wall}}{T_{G\infty} - T_{wall}}$ [–]
0.00					0	0.0000	
0.005	159.89	3.88	156.0	6.47E-02	4.17E-04	0.0192	0.170
0.01	113.37	2.85	110.5	4.57E-02	6.93E-04	0.0287	0.136
0.02	80.32	2.07	78.2	3.23E-02	1.08E-03	0.0422	0.109
0.04	56.82	1.5	55.3	2.28E-02	1.62E-03	0.0610	0.087
0.06	46.43	1.24	45.2	1.86E-04	2.03E-03	0.0753	0.076
0.08	40.23	1.08	39.1	1.61E-02	2.38E-03	0.0873	0.0690
0.10	35.99	0.97	35.0	1.44E-02	2.68E-03	0.0979	0.064
0.12	32.83	0.89	32.0	1.31E-02	2.96E-03	0.1075	0.061
0.14	30.40	0.82	29.6	1.22E-02	3.21E-03	0.1163	0.058
0.16	28.45	0.77	27.7	1.14E-02	3.45E-03	0.1245	0.055
0.18	26.80	0.73	26.1	1.07E-02	3.67E-03	0.1322	0.053
0.20	25.44	0.69	24.8	1.02E-02	3.88E-03	0.1395	0.051

Step 5: Comparison with Nusselt's model

For comparison purposes, the theoretical value derived from Nusselt's model is calculated by assuming $T_s = T_{G\infty}$:

$T_s = T_G = 371.72$ K

$T_{ref} = 298.15 + 0.31(371.72 - 298.15) = 320.96$ K

$\lambda_s = 2270$ kJ kg^{-1}

$\lambda_c = 2270 + (3/8)(4.18)(371.72 - 298.15) = 2.38 \times 10^3$ kJ kg^{-1},

$\rho_L = 988.85$ kg m^{-3}, $\kappa_L = 0.639$ W m^{-1} K^{-1}, $\mu_L = 0.5685$ mPa s,

$\rho_G = 0.61$ kg m^{-3}

$h_c = (0.943)\left\{ \dfrac{(9.81)(988.85)(988.85 - 0.61)(0.639)^3(2.38 \times 10^6)}{(0.5685 \times 10^{-3})(372.72 - 298.15)(0.2)} \right\}^{1/4}$

$= 4.87$ kW m^{-2} K^{-1}

$Q_{Nusselt} = (4.87)(0.2)(371.72 - 298.15) = 71.67$ kW m^{-1}

$Q/Q_{Nusselt} = (10.0)/(71.67) = 0.1395$

The calculation indicates that the condensation heat transfer rate reduces to only 14% of the theoretical rate based on the Nusselt model, with the assumption that $T_s = T_{G\infty}$. This may be due to the fact that the transport phenomena in the case of condensation in the presence of a non-condensable gas are vapor-phase

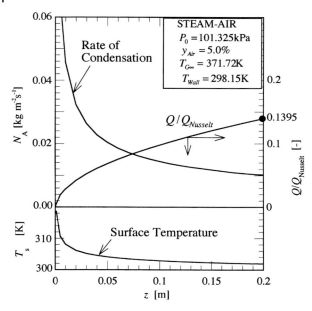

Fig. 9.10 Condensation of steam with 5% air on a vertical flat plate at a total pressure of 1 atm (forced-flow conditions) in Example 9.4.

mass transfer controlled, whereas the condensation of pure vapors is liquid-phase heat transfer controlled. In the present case, the relative order of magnitude of the liquid-phase heat transfer resistance is about 7% compared to the total resistance and the relative order of magnitude of the vapor-phase sensible heat flux, q_G, compared to the vapor-phase latent heat flux, λN_A, is about 3%.

Figure 9.10 shows the variation in the interface temperature, T_s, the rate of heat transfer, $Q/Q_{Nusselt}$, and the vapor-phase local mass flux, N_A. The interface temperature decreases rapidly near the leading edge of the wall, as does the local rate of condensation. This indicates that the contribution of the upper portion of the condensing wall is dominant while that of the lower portion is much less important.

9.6
Condensation of Vapors on a Circular Cylinder

9.6.1
Condensation of Pure Vapors on a Horizontal Cylinder

The discussions in the previous sections have been confined to condensation on a vertical flat plate. In practical applications, however, horizontal tubes are used as condensing surfaces rather than flat plates. In this section, we discuss condensation on a short horizontal tube, where the axial variation of the wall temperature or that of the cooling water inside the condensing tube is negligibly small.

The average condensation heat transfer coefficient over the surface of a cylinder is given by the following equation, the derivation of which is similar to that of Eq. (9.20):

$$\overline{h_c} = 0.725 \left\{ \frac{g\rho_L(\rho_L - \rho_G)\kappa_L^3 \lambda_c}{\mu_L(T_s - T_{wall})D_T} \right\}^{1/4} \tag{9.28}$$

The average liquid-phase sensible heat flux over the surface of the condensing tube is given by the following equation:

$$\overline{q_L} = \overline{h_c}(T_s - T_{wall}) \tag{9.29}$$

from which the rate of condensation of vapor per unit length of the condensing tube, Γ [kg m^{-1} s^{-1}] is given by:

$$\Gamma = \pi D \overline{q_L}/\lambda_s \tag{9.30}$$

9.6.2
Heat and Mass Transfer in the case of a Cylinder with Surface Mass Injection or Suction

As discussed in the previous sections, the vapor-phase sensible heat and mass fluxes of a cylinder are fundamental to the prediction of the rates of condensation of vapors in the presence of a non-condensable gas on a horizontal tube. M. W. Chang and B. A. Finlayson [3] described a numerical approach to heat transfer for a cylinder in the case of no surface mass suction. They proposed the following correlation:

$$\overline{Nu_0} = (0.36 + 0.58 \, Re^{0.48}) \, Pr^m \tag{9.31}$$

$$m = 0.29 + 0.028 \log Re \tag{9.32}$$

where Re is the Reynolds number ($Re \equiv \rho_G D_T U/\mu_G$).

T. Mamyoda and K. Asano [5] also presented a numerical approach to heat and mass transfer in the case of a cylinder with surface mass injection or suction for Reynolds numbers in the range 1 to 100 and Schmidt numbers in the range 0.5 to 2.0. They proposed the following correlation for the average diffusional flux over the surface of a cylinder in the case of no mass injection or suction:

$$\overline{Nu_0} = 0.66 \, Re^{0.46} \, Pr^{0.36} \tag{9.33}$$

$$\overline{Sh_0}(1 - \omega_s) = 0.66 \, Re^{0.46} \, Sc^{0.36} \tag{9.34}$$

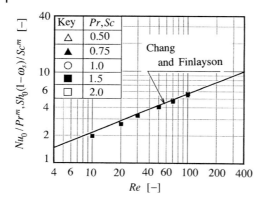

Fig. 9.11 Heat transfer for a cylinder; comparison of numerical data obtained by T. Mamyoda and K. Asano [5] with the correlation by M. W. Chang and B. A. Finlayson [3].

Figure 9.11 shows a comparison of their numerical data with Chang and Finlayson's correlation, which shows good agreement. They also proposed correlations for the local heat and diffusional fluxes at $\theta = 0$, $\pi/2$, $2\pi/3$, $5\pi/6$, and π, based on the following equations:

$$Nu_{\theta,0} = a + bRe^m Pr^{0.36} \tag{9.35}$$

$$Sh_{\theta,0}(1 - \omega_s) = a + bRe^m Sc^{0.36} \tag{9.36}$$

from the values of which the local distribution of the fluxes on the surface of a cylinder can be estimated by an interpolation method [6].

Table 9.5 summarizes the coefficients of Eqs. (9.35) and (9.36).

Table 9.5 Coefficients of Eqs. (9.35) and (9.36) [6].

θ	a	b	m
0	0	0.91	0.53
$(1/2)\pi$	0	0.82	0.39
$(2/3)\pi$	0.57	0.53	0.23
$(5/6)\pi$	0.50	0.09	0.61
π	0.41	0.07	0.74

On the basis of their numerical data, they proposed the following correlation for the effect of mass injection or suction on heat and mass transfer in the case of a cylinder [5]:

For heat transfer:

$$g(B_H) \equiv \frac{\overline{Nu}}{\overline{Nu_0}} = \frac{1}{c + (1-c)(1+B_H)^{0.67}} \tag{9.37}$$

$$c = 0.07 + 0.16 \log Pr \tag{9.38}$$

For mass transfer:

$$g(B_M) \equiv \frac{\overline{Sh}(1-\omega_s)}{\overline{Sh_0}(1-\omega_s)} = \frac{1}{c + (1-c)(1+B_M)^{0.67}} \tag{9.39}$$

$$c = 0.07 + 0.16 \log Sc \tag{9.40}$$

where B_H is the transfer number for heat transfer, and B_M is the transfer number for mass transfer.

9.6.3
Calculation of the Rates of Condensation of Vapors on a Horizontal Tube in the Presence of a Non-Condensable Gas

The rates of condensation of vapors on a horizontal tube in the presence of a non-condensable gas under forced-flow conditions can be calculated in a similar way as in the case of a vertical flat plate, where the interface temperature is determined from the heat balance at the interface.

Figure 9.12 shows the effect of a non-condensable gas on the rate of condensation of vapors on a short horizontal tube under forced-flow conditions, as determined by T. Mamyoda and K. Asano [6]. The ordinate represents the condensation heat flux normalized by the theoretical value according to Nusselt's model with the assumption of $T_s = T_{G\infty}$ in Eq. (9.28). The abscissa is the mass fraction of air. The three solid lines represent the results of numerical simulations, based on Eq. (9.28) for the liquid-phase heat flux and Eqs. (9.35)–(9.40), where the local interface temperatures are determined from the energy balance at the interface. The effect of a non-condensable gas can be seen to be less pronounced than in Othmer's experiment. This may be due to the fact that Othmer's experiment was conducted under natural convection, whereas Mamyoda and Asano employed forced-flow conditions, under which there is less accumulation of the non-condensable gas.

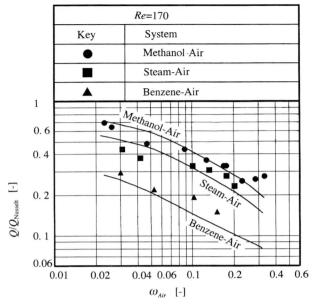

Fig. 9.12 Condensation of vapors on a short horizontal tube in the presence of a non-condensable gas; comparison of observed data with simulation [6].

Example 9.5

Saturated steam with 5% air condenses on the surface of a horizontal tube of diameter 20 mm. The steam is supplied vertically downward at a velocity of 0.2 m s^{-1} and the wall temperature of the tube is 298.15 K. Calculate the rate of condensation per unit length of the tube [kg m^{-1} s^{-1}].

Solution

The physical properties of the system are the same as in Example 9.4. Here, we use the following physical properties:

$T_G = 371.72$ K, $\rho_G = 0.609$ kg m^{-3}, $\kappa_{Gs} = 2.50 \times 10^{-2}$ W m^{-1} s^{-1},

$D_G = (2.56 \times 10^{-5}) T/298.15)^{1.83}$ m^2 s^{-1}, $Pr = 0.76$, $Sc = 0.61$

We shall apply a rather approximate method using the average fluxes instead of local fluxes. The principle of the calculation is to find a suitable interface temperature of the liquid by a *trial-and-error method* that will satisfy the following equation:

$$\overline{q_L} = \overline{N_A}\lambda_s + \overline{q_G} \tag{A}$$

where $\overline{q_L}$ is the average sensible heat flux in the liquid phase estimated by Eq. (9.28), $\overline{N_A}$ is the average vapor-phase mass flux according to Eqs. (9.34) and

(9.39), and $\overline{q_G}$ is the average vapor-phase sensible heat flux according to Eqs. (9.33) and (9.37) by assuming $B_H = B_M$.

1) First trial

As an initial value for the calculation, we tentatively assume T_s = 320 K.

Step 1: Calculation of liquid-phase sensible heat flux:

Reference temperature for estimation of the physical properties of the liquid:

$$T_{ref} = 298.15 + 0.31(320-298.15) = 304.92 \text{ K}$$

The physical properties of the liquid at the reference temperature are as follows:

$\rho_L = 994.2$ kg m^{-3}, $\kappa_L = 0.619$ W m^{-1} K^{-1}, $\mu_L = 0.767$ mPa s,

$\lambda_s = 2.39 \times 10^6$ J kg^{-1}

$\lambda_c = 2.39 \times 10^6 + (3/8)(4.18 \times 10^3)(320-298.15) = 2.42 \times 10^6$ J kg^{-1}

The average condensation heat transfer coefficient is given by Eq. (9.28):

$$\overline{h_c} = 0.725 \left\{ \frac{(9.81)(994.2)^2(0.619)^3(2.42 \times 10^6)}{(0.763 \times 10^{-3})(320 - 298.15)(0.02)} \right\}^{1/4}$$

$$= 8.23 \times 10^3 \text{ W m}^{-2} \text{ K}^{-1}$$

The average sensible heat flux of the liquid:

$$\overline{q_L} = (8.23 \times 10^3)(320 - 298.15)/(1000) = 179.9 \text{ kW m}^{-2}$$

Step 2: Calculation of vapor-phase sensible heat flux and latent heat flux:

$\log p_s = 10.0917 - 1668.21/(320 - 45.15)$

$p_s = 10.524$ kPa, $y_s = 0.1039$, $\omega_s = 0.0672$,

$B = (0.0672 - 0.9220)/(1 - 0.0672) = -0.916$

Calculation of $\overline{Nu_0}$ and $\overline{Sh_0}(1-\omega_s)$

$Re = (0.5)(0.609)(0.02)/(1.95 \times 10^{-5}) = 312.3$

$\overline{Nu_0} = (0.66)(312.3)^{0.46}(0.76)^{0.36} = 8.40$

$\overline{Sh_0}(1-\omega_s) = (0.66)(312.3)^{0.46}(0.61)^{0.36} = 7.76$

The effect of high mass flux due to condensation of vapor:

$c = 0.07 + 0.16 \log(0.61) = 0.0356$

$g(B) = 1/\{0.0356 + (1 - 0.0356)(1 - 0.919)^{0.67}\} = 4.57$

Calculation of sensible heat flux and latent heat flux:

$$\overline{q_G} = (8.40)(4.57)(0.0250)(371.72 - 320)/(0.02)/1000 = 2.48 \text{ kW m}^{-2}$$

$$\overline{N_A} = (7.76)(4.57)(0.916)(1.060)(2.91 \times 10^{-5})/(0.02) = 5.023 \times 10^{-2} \text{ kg m}^{-2} \text{ s}^{-1}$$

$$\overline{N_A}\lambda_s = (5.023 \times 10^{-2})(2.39 \times 10^3) = 120.1 \text{ kW m}^{-2}$$

Step 3: Heat balance at the interface:

$$(\overline{q_L} - \overline{N_A}\lambda_s - \overline{q_G})/\overline{q_L} = (179.9 - 120.1 - 2.48)/(179.9) = 0.319$$

The assumed interface temperature does not satisfy Eq. (A).

Similar calculations are repeated until the assumed interface temperature satisfies the heat balance at the interface. Table 9.6 shows the results of such calculations.

Table 9.6 Calculation of the rate of condensation of steam with 5% air on a horizontal tube at 1 atm in Example 9.5.

	T_s [K]	*371.72	320	310	311.9	311.91
Liquid phase heat flux	T_{ref} [K]	320.96	304.92	301.82	302.41	302.42
	ρ_L [kg m^{-3}]	988.9	994.2	995.3	995.1	995.1
	κ_L [W m^{-1} K^{-1}]	6.39E-01	6.19E-01	6.15E-01	6.15E-01	6.15E-01
	μ_L [mPa s]	0.5685	0.7673	0.8195	0.8091	0.8091
	λ_s [J kg^{-1}]	2.27E+06	2.39E+06	2.41E+06	2.41E+06	2.41E+06
	λ_c [J kg^{-1}]	2.38E+06	2.42E+06	2.43E+06	2.43E+06	2.43E+06
	h_c [W m^{-2} K^{-1}]	6.66E+03	8.23E+03	9.40E+03	9.09E+03	9.09E+03
	q_L [kW m^{-2}]	490.23	179.85	111.38	125.01	125.08
	Γ [kg m^{-1} s^{-1}]	1.36E-02	4.72E-03	2.90E-03	3.26E-03	3.26E-03
Vapor phase fluxes	Nu_0		8.40	8.40	8.40	8.40
	$Sh_0 (1 - \omega_s)$		7.76	7.76	7.76	7.76
	p_s [kPa]		10.52	6.21	6.88	6.89
	B		−0.916	−0.919	−0.918	−0.918
	$g(B)$		4.575	4.652	4.641	4.640
	ρ_G [kg m^{-3}]		1.060	1.113	1.103	1.103
	$D(T_s)$ [m^2 s^{-1}]		2.91E-05	2.75E-05	2.78E-05	2.78E-05
	N_A [kg m^{-2} s^{-1}]		5.03E-02	5.07E-02	5.07E-02	5.07E-02
	$N_A \lambda_s$ [kW m^{-2}]		1.20E+02	1.22E+02	1.22E+02	1.22E+02
	q_G [kW m^{-2}]		2.48E+00	3.01E+00	2.91E+00	2.91E+00
	Heat balance $1 - q_G/q_L - N_A\lambda_s/q_L$		0.3183	−0.1268	−0.0009	−0.0003

* Theoretical value by the Nusselt model

From Tab. 9.6, we obtain the following results:

$T_s = 311.91$ K

$\overline{h_c} = 9.09 \times 10^3$ W m^{-2}K^{-1}

$\overline{q_L} = 9.09\,(311.91 - 298.15) = 125.1$ kW m^{-2}

$\overline{q_G} = 2.91$ kW m^{-2}

$\overline{N_A} = 5.07 \times 10^{-2}$ kg m^{-2}s^{-1}

$\overline{N_A}\lambda_s = 122.1$ kW m^{-2}

$(\overline{q_L} - \overline{N_A}\lambda_s - \overline{q_G})/\overline{q_L} = (125.1 - 122.1 - 2.91)/(125.1) = 0.0002$

The rate of condensation per unit length of the tube is given by:

$\Gamma = (3.14)(0.02)(5.07 \times 10^{-2}) = 3.18 \times 10^{-3}$ kg m^{-1} s^{-1}

The theoretical value based on the Nusselt model is given by substituting $T_s = T_G = 371.72$ K into Eq. (9.26):

$$\overline{h_{c\,Nusselt}} = 0.725\left\{\frac{(9.81)(988.9)^2(0.639)^3(2.38 \times 10^6)}{(0.5685 \times 10^{-3})(371.72 - 298.15)(0.02)}\right\}^{1/4}$$

$= 6.66 \times 10^3$ W m^{-2} K^{-1}

$\overline{q_{L\,Nusselt}} = (6.66 \times 10^3)(371.72 - 298.15)/(1000) = 490.23$ kW m^{-2}

$\Gamma_{Nusselt} = \dfrac{h_c(T_s - T_{wall})(\pi D_T)}{\lambda_s} = 1.36 \times 10^{-2}$ kg m^{-1}s^{-1}

$\Gamma/\Gamma_{Nusselt} = (3.18 \times 10^{-3})/(1.36 \times 10^{-2}) = 0.240$

The rate of condensation in the presence of a non-condensable gas is reduced to 24 % of the theoretical value by the Nusselt model.

References

1 K. Asano, Y. Nakano, and M. Inaba, "Forced Convection Film Condensation of Vapors in the Presence of Noncondensable Gas on a Small Vertical Flat Plate", *Journal of Chemical Engineering of Japan*, **12**, [3], 196–202 (1979).

2 K. Asano and S. Matsuda, "Total and Partial Condensation of Binary Vapor Mixtures of Methanol/Water Systems on a Vertical Flat Plate", *Kagaku Kogaku Ronbunshu*, **11**, [1], 27–33 (1985).

3 M. W. Chang and B. A. Finlayson, "Heat Transfer in Flow-past Cylinders at $Re < 150$; Part 1. Calculations for Constant Fluid Properties", *Numerical Heat Transfer*, **12**, 179–195 (1987).

4 A. P. Colburn and T. B. Drew, "The Condensation of Mixed Vapors", *Transactions of the American Institute of Chemical Engineers*, **33**, 197–211 (1937).

5 T. Mamyoda and K. Asano, "Numerical Analysis of Drag Coefficients and Heat and Mass Transfer of a Cylinder with Diffusive Surface Mass Suction or Injection", *Journal of Chemical Engineering of Japan*, **26**, [4], 382–388 (1993).

6 T. Mamyoda and K. Asano, "Experimental Study of Condensation of Vapors in the Presence of Noncondensable Gas on a Short Horizontal Tube", *Journal of Chemical Engineering of Japan*, **27**, [4], 485–491 (1994).

7 W. J. Minkowycz and E. M. Sparrow, "Condensation Heat Transfer in the Presence of Non-condensables, Interfacial Resistance, Superheating, Variable Properties, and Diffusion", *International Journal of Heat and Mass Transfer*, **9**, 1125–1144 (1966).

8 T. Mizushina and M. Nakajima, "Simultaneous Heat and Mass Transfer; The Relation between Heat Transfer and Material Transfer Coefficients", *Kagaku Kikai*, **15**, [1], 30–34 (1951).

9 W. von Nusselt, "Die Oberfrächenkondensation des Wasserdampfes", *Zeitschrift Vereines Deutscher Ingenieure*, **60**, [27], 541–546 (1916).

10 D. F. Othmer, "Condensation of Steam", *Industrial and Engineering Chemistry*, **21**, [6], 576–583 (1929).

11 T. K. Sherwood and E. W. Comings, "An Experimental Study of the Wet-Bulb Hygrometer", *Transactions of the American Institute of Chemical Engineers*, **28**, [1], 88–117 (1932).

12 E. M. Sparrow, W. J. Minkowycz, and M. Saddy, "Forced Convection Condensation in the Presence of Non-condensables and Interfacial Resistance", *International Journal of Heat and Mass Transfer*, **10**, 1829–1845 (1967).

10
Mass Transfer in Distillation

10.1
Classical Approaches to Distillation and their Paradox

10.1.1
Tray Towers and Packed Columns

Distillation is one of the most stable, reliable, and general-purpose separation techniques for the separation of liquid mixtures, and because of this it is widely used in almost all fields of the chemical industry. Indeed, more than 90% of all separation processes are carried out in distillation plants. Although classical approaches to distillation offer some guidelines for the design of actual separation processes, the conventional methods are inadequate for addressing the requirements that have arisen from the recent trends for fine separation and reducing energy consumption of the processes. In this chapter, we discuss the possibilities of a new approach to distillation on the basis of the *Simultaneous Heat and Mass Transfer model*, the details of which are discussed in the subsequent sections.

The two major types of distillation apparatus for industrial use are tray towers and packed columns, which are based on completely different design principles. Thus, while the former are based on an equilibrium stage model modified with tray efficiencies, the latter are based on a rate model with the parameter HTU (height per transfer unit), similar to the treatment of gas absorption. Tray towers are widely used for industrial separation processes, especially in large capacity plants, largely as a result of their simple construction, stability, and ease of operation. However, the recent development of new and highly efficient packings has led to a reevaluation of the merits of packed distillation columns, especially with regard to low energy consumption, aptitude for fine separation, and the possibility of building more compact plants. For this reason, the focus of this chapter is on packed distillation columns.

Mass Transfer. From Fundamentals to Modern Industrial Applications. Koichi Asano
Copyright © 2006 WILEY-VCH Verlag GmbH & Co. KGaA, Weinheim
ISBN: 3-527-31460-1

10.1.2
Tray Efficiencies in Distillation Columns

The conventional method for the design of tray towers is based on calculation of the number of equilibrium stages necessary for the specified separation criteria, with a correction for tray efficiencies.

Figure 10.1 shows a schematic representation of mass transfer on a tray in a distillation column. In 1925, E. V. Murphree [17] presented a theoretical approach to mass transfer in tray tower distillation columns, for which the following assumptions were made:

1. Liquid concentrations on the tray are uniform.
2. The rate of mass transfer of each component is proportional to its concentration driving force.

The material balance for a small volume element of the liquid, dz, on the *n-th* tray from the bottom of the column, can be expressed by the following equation:

$$G_M dy_i = K_i a(y_{i,n}^* - y_i)dz \tag{10.1}$$

The boundary conditions are:

$$z = 0: \quad y_i = y_{i,n-1}$$
$$z = Z_c: \quad y_i = y_{i,n}$$

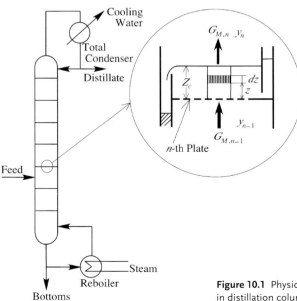

Figure 10.1 Physical picture of tray efficiencies in distillation columns

Integrating Eq. (10.1) with respect to z and rearranging the resulting equation, we obtain the following equation:

$$E_{MV, i, n} \equiv \frac{y_{i,n} - y_{i,n-1}}{y^*_{i,n} - y_{i,n-1}} = 1 - \exp(-K_i a Z_c / G_M) \tag{10.2}$$

where a is the interfacial area for mass transfer per unit volume of the liquid [m^{-1}], G_M is the molar velocity of the vapor [kmol m^{-2} s^{-1}], K_i is the mass transfer coefficient of component i [kmol m^{-2} s^{-1}], $y_{i,n-1}$ and $y_{i,n}$ are the mole fractions of component i of the vapor leaving from the $(n-1)$-th and the n-th tray, respectively, $y^*_{i,n}$ is the mole fraction of the vapor in equilibrium with the liquid on the n-th tray, and Z_c is the clear liquid height on the tray [m].

The term $E_{MV,i,n}$ on the left-hand side of Eq. (10.2) is known as the *Murphree's (vapor-phase) tray efficiency* of the component i on the n-th tray, which is defined as the ratio of the actual change in the vapor concentration on the n-th tray to that on the ideal stage (equilibrium stage), and represents a measure of the rate of mass transfer on the tray.

10.1.3
HTU as a Measure of Mass Transfer in Packed Distillation Columns

In 1935, T. H. Chilton and A. P. Colburn [4] proposed their famous concept of HTU for estimation of the packed height necessary to achieve specified separation criteria in gas absorption by packed columns. The basic assumption regarding HTU is that the mass transfer of a given component is proportional to its concentration driving force. The authors applied the following equation for mass transfer in a small volume element of the packed column, dz:

$$G_M dy_i = K_i a (y^*_i - y_i) dz \tag{10.3}$$

Integrating Eq. (10.3) with respect to packed height z, the packed height necessary for the specified separation criteria, Z_T, is given by the following equation:

$$Z_T = \left(\frac{G_M}{K_i a}\right) \int_{y_{i,\text{Bottom}}}^{y_{i,\text{Top}}} \frac{1}{(y^*_i - y_i)} dy_i \tag{10.4}$$

where the term $(G_M/K_i a)$ on the right-hand side of Eq. (10.4) is called the HTU (height per transfer unit) [m] and represents a measure of the rate of mass transfer in the column. The integral on the right-hand side of Eq. (10.4) is called the *number of transfer units* for component i, which represents a measure of the mass transfer resistances for the specified separation.

$$\text{Number of transfer units} \equiv \int_{y_{i,\,Bottom}}^{y_{i,\,Top}} \frac{1}{(y_i^* - y_i)} \, dy_i$$

Equation (10.4) indicates that the packed height necessary to meet the specified separation criteria is given as the product of HTU and the number of transfer units.

10.1.4
Paradox in Tray Efficiency and HTU

The theoretical basis of classical approaches to distillation is that distillation is a simple mass transfer process, as in the case of gas absorption, and that the rates of mass transfer of components are proportional to their concentration driving forces. The study of tray efficiencies in distillation columns by the *AIChE Research Committee* [1] in the late 1950s is a typical example of the classical approach. Correlations were proposed for the estimation of tray efficiencies based on wide ranges of observed data for gas absorption in a pilot-plant-scale tray tower. However, later studies suggested that the results obtained by these classical approaches are not applicable to ordinary distillation.

G. G. Haselden and R. M. Thorogood [9] reported that tray efficiencies in the cryogenic distillation of air showed a considerable dependence on the liquid concentrations on the tray. A similar dependence on the liquid concentration was reported by M. W. Bidulph and M. M. Dribika [3] for binary distillations of the methanol/propanol and ethanol/propanol systems.

Figure 10.2 shows observed tray efficiency data relating to ternary distillation of the acetone/methanol/water system in bubble-cap tray towers obtained by A. Vo-

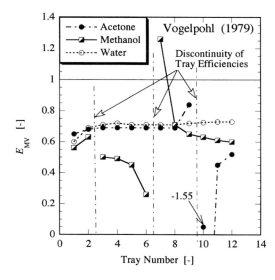

Figure 10.2 Discontinuity of tray efficiencies in ternary distillation of the acetone/methanol/water system; data from A. Vogelpohl. [20]

gelpohl [20]. The figure clearly indicates that the tray efficiencies of the various components show quite different values. Moreover, the tray efficiencies of acetone and methanol show curious discontinuities in the middle of the column, where E_{MV} exceeds 100 %, and then they suddenly take negative values, for which the classical model cannot give any reasonable explanation. If the mechanisms of mass transfer in gas absorption and in distillation are the same, the tray efficiencies of distillation columns should be independent of the liquid concentrations on the tray. Moreover, from the condition of positive finite values of mass transfer coefficients, Eq. (10.2) indicates that the tray efficiencies in distillation columns must satisfy the following condition:

$$0 < E_{MV,i,n} < 1 \tag{10.6}$$

However, Vogelpohl's data indicate that Eq. (10.6) does not hold for distillation.

Figure 10.3 shows observed data for ternary distillation of the acetone/methanol/ethanol system in a wetted-wall column, as obtained by H. Kosuge et al. [13] In the figure, a series of data along a distillation path, where the bottom concentrations of the successive measurements are adjusted to be nearly equal to the top concentrations of the previous ones, is shown as a function of the liquid concentration of methanol. Discontinuities in the number of transfer units for methanol are also observed near the region where the distillation path intersects the zero driving force line for methanol. This fact is inconsistent with the conclusions of the classical model, in which the rates of mass transfer are assumed to be propor-

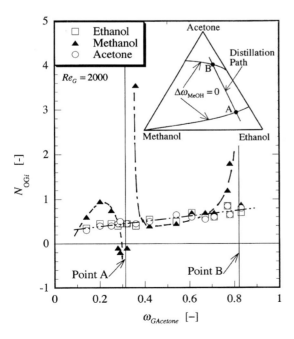

Figure 10.3 Discontinuities in the number of transfer units of the intermediate component in ternary distillation on a wetted-wall column; data from H. Kosuge et al. [13]

tional to the corresponding concentration driving forces, and the distillation path cannot intersect the zero driving force line, at which the mass transfer resistances become infinite. Similar discontinuities of the HTU have been reported by H. Kosuge et al. [15] for ternary distillation of the acetone/methanol/ethanol system on a packed distillation column and by N. Egoshi et al. [5] for cryogenic distillation of the nitrogen/argon/oxygen system on a wetted-wall column. These experimental findings suggest that there must be something wrong with the theoretical foundations of the classical model for distillation.

Example 10.1
The following data are obtained for binary distillation of the acetone/ethanol system on a 200 mm packed height distillation column at 1 atm under total reflux conditions.

Concentration of the top liquid: $x_D = 0.56$
Concentration of the bottom liquid: $x_B = 0.02$

Calculate the liquid concentration at the top of the column for the same bottom concentration and under the same operating conditions if the packed height is increased to 600 mm.

Solution
As will become apparent in the later discussions, an approach based on the *Simultaneous Heat and Mass Transfer* model is preferable. However, here we will apply an approximate solution based on the classical model with HTU.

Step 1. Estimation of the vapor-liquid equilibrium:
The vapor-liquid equilibrium (VLE) of the system is estimated from the vapor pressures of the pure components derived from Antoine's equation and the activity coefficients derived from Wilson's equation. Readers interested in the details of the VLE calculations may refer to standard textbooks on chemical engineering thermodynamics. Table 10.1 summarizes the results of the VLE calculation. The first column represents liquid concentration and the second column equilibrium vapor concentration.

Step 2. Calculation of the number of transfer units
Since the distillation is under total reflux conditions, we have the following equation for the operating line from the material balance:

$$y = x$$

From the definition of the number of transfer units, we have:

$$N_{oy} = \int_{0.02}^{y} \frac{dy}{y^* - y}$$

Table 10.1 Vapor-liquid equilibrium and the numbers of transfer units for the acetone/ethanol system under total reflux conditions at 1 atm (Example 10.1).

$x = y$	y^*	$y^* - y$	$\int_{0.02}^{y} \frac{1}{(y^*-y)} dy$
0.02	0.0690	0.0490	0.000
0.04	0.1303	0.0903	0.315
0.06	0.1850	0.1250	0.506
0.08	0.2340	0.1540	0.651
0.10	0.2774	0.1774	0.772
0.15	0.3714	0.2214	1.026
0.20	0.4461	0.2461	1.240
0.25	0.5073	0.2573	1.439
0.30	0.5587	0.2587	1.633
0.35	0.6029	0.2529	1.828
0.40	0.6416	0.2416	2.031
0.45	0.6761	0.2261	2.245
0.50	0.7079	0.2079	2.475
0.55	0.7373	0.1873	2.729
0.60	0.7654	0.1654	3.014
0.65	0.7923	0.1423	3.341
0.70	0.8192	0.1192	3.726
0.75	0.8460	0.0960	4.196
0.80	0.8733	0.0733	4.798
0.85	0.9016	0.0516	5.623
0.90	0.9320	0.0320	6.889
0.92	0.9446	0.0246	7.608
0.94	0.9577	0.0177	8.579
0.96	0.9714	0.0114	10.022

The fourth column in Tab. 10.1 contains numerical values for the numbers of transfer units corresponding to vapor concentration y.

Step 3. Calculation of HTU
From Tab. 10.1, we have the following values:

$x = 0.55, \quad N_{oy} = 2.729$

$x = 0.60, \quad N_{oy} = 3.014$

Interpolating the above values, for $x_D = 0.56$ we obtain the following value for N_{oy}:

$N_{oy} = 2.78$

From Eq. (10.4), HTU under the given operating conditions is estimated according to:

HTU = (0.20)/(2.78) = 0.0719 m

Step 4. Calculation of the liquid concentration at the top of the column

Since the numerical value of HTU is the same under the same operating conditions, the number of transfer units for the 600 mm column is given by:

$N_{oy} = (0.6)/(0.0719) = 8.34$

From Tab. 10.1, we have:

$x = 0.92, \quad N_{oy} = 7.608$

$x = 0.94, \quad N_{oy} = 8.579$

Interpolating the above values, for N_{oy} we obtain the following value for x_D:

$x_D = 0.935$

The observed liquid concentration corresponding to the given operating conditions has been reported by H. Kosuge et al. [14] as:

$x_{D,ob} = 0.95$

In the case of such a binary distillation, agreement between the data and the approximate calculation by HTU is reasonable.

10.2
Characteristics of Heat and Mass Transfer in Distillation

10.2.1
Physical Picture of Heat and Mass Transfer in Distillation

Figure 10.4 shows a physical picture of heat and mass transfer in binary distillation. In the distillation column, the mixture of vapors is in contact with the liquid and transfer of the following physical quantities, which interact through the boundary conditions, takes place at the vapor-liquid interface.

1. Sensible heat flux due to the temperature difference between the bulk vapor and the interface, q_G [W m^{-2}].
2. Sensible heat flux due to the temperature difference between the interface and the bulk liquid, q_L [W m^{-2}].
3. Vapor-phase diffusional flux due to the concentration difference between the vapor at the interface and the bulk vapor, $J_{G,i,s}$ [kg m^{-2} s^{-1}].
4. Liquid-phase diffusional flux due to the concentration difference between the bulk liquid and the liquid at the interface, $J_{L,i,s}$ [kg m^{-2} s^{-1}].
5. Convective mass fluxes of each component, $\rho_s v_s \omega_{i,s}$ [kg m^{-2} s^{-1}].
6. Latent heat flux of each component due to the phase change, $\lambda_i N_i$ [W m^{-2}].

The principle of local equilibrium stipulates that the vapor is in equilibrium with the liquid at the interface.

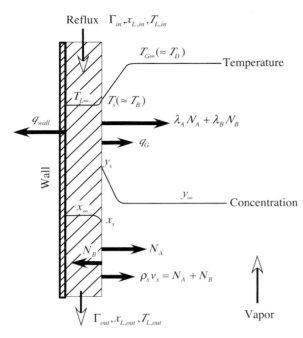

Figure 10.4 Physical picture of heat and mass transfer in binary distillation

From the condition of no accumulation of energy at the interface, we have the following equation:

$$\lambda_A N_A + \lambda_B N_B + q_G + q_w = 0 \tag{10.7}$$

In the following, we discuss the characteristics of heat and mass transfer in distillation.

10.2.2
Rate-Controlling Process in Distillation

Figure 10.5 shows the observed vapor-phase temperature distribution for binary distillation of the ethanol/water system on a flat-plate wetted-wall column, as obtained by A. Ito and K. Asano [10]. The solid line in the figure represents the theoretical temperature distribution based on laminar boundary theory with surface mass suction, and shows good agreement with the observed data. The observed bulk vapor temperature, $T_{G\infty}$, agrees well with the dew point of the vapor, and the observed bulk liquid temperature, $T_{L\infty}$, is consistent with the bubble point of the liquid. The observed interface temperature, $T_{s,\,ob}$ (= 357.5 K), which is obtained by extrapolating the vapor-phase temperature distribution to the interface, is slightly higher than the bubble point temperature of the liquid, $T_B(x_{bulk})$ (= 356.6 K).

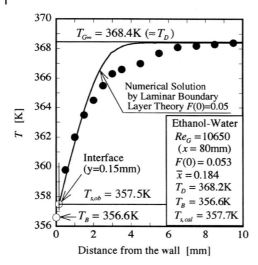

Figure 10.5 Vapor-phase temperature distribution in binary distillation of the ethanol/water system on a flat-plate wetted-wall column; data from A. Ito and K. Asano [10]

The bubble point of the liquid, corresponding to the liquid interface concentration, x_s, is estimated from the bulk liquid concentration by the use of Higbie's penetration model, as described in Section 3.4.2:

$$k_c = 2\sqrt{(D_L/\pi)(L/u_{Ls})} \tag{10.8}$$

$$x_s = x_{bulk} - (N_A/M_A)/ck_c \tag{10.9}$$

$T_B(x_s)$ (= 357.5 K) agrees well with the observed interface temperature within experimental error (± 0.2 K). Similar results have also been obtained for the methanol/water system, leading to the conclusion that more than 90% of the thermal resistances are in the vapor phase, and the same is true for mass transfer. H. Kosuge and K. Asano [12] also presented a similar approach to the vapor-phase temperature distribution for ternary distillation of the methanol/ethanol/water system on a wetted-wall column, and reported that this system is also vapor-phase controlled.

10.2.3
Effect of Partial Condensation of Vapors on the Rates of Mass Transfer in Binary Distillation

Figure 10.6 shows data for binary distillation of the methanol/water system on a flat-plate wetted-wall column under non-adiabatic conditions, as obtained by A. Ito and K. Asano [11]. The figure indicates that the mass flux of methanol, the more volatile component, decreases with increasing amount of partially condensed vapor, and even takes a negative value when the amount of partially condensed vapor exceeds a certain critical value. This negative mass flux of methanol means that the transfer of the methanol takes place in the opposite direction to

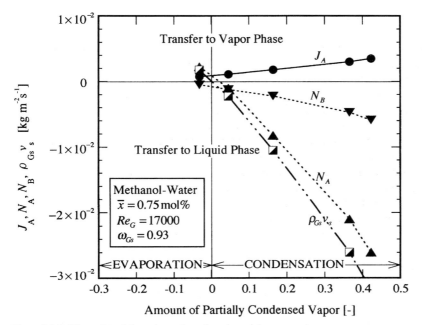

Figure 10.6 Effect of partial condensation of vapors on the vapor-phase mass and diffusional fluxes in binary distillation of the methanol/water system; data from A. Ito and K. Asano [11].

that expected from the concentration driving force, i.e. from the vapor phase to the interface.

This paradox may be rationalized by the following reasoning. The mass flux may be expressed as the sum of the diffusional flux and the convective mass flux, as discussed in Section 2.3.1:

$$N_A = J_{A,s} + \rho_s v_s \omega_{A,s} \tag{10.10a}$$

$$N_B = J_{B,s} + \rho_s v_s \omega_{B,s} \tag{10.10b}$$

The convective mass flux for mass transfer in binary distillation is given by Eq. (2.37), as discussed in Section 2.3.4.

$$\rho_s v_s = \frac{(\lambda_{B,s} - \lambda_{A,s}) J_{A,s} - q_G - q_w}{\lambda_{A,s} \omega_{A,s} + \lambda_{B,s} \omega_{B,s}} \tag{2.37}$$

where the subscript 's' denotes conditions at the interface. Equation (2.37) indicates that the convective mass flux, $\rho_s v_s$, decreases with increasing q_w, and even takes a negative value if q_w exceeds a certain critical value, and so the mass flux also decreases. However, even in this region of negative mass flux of methanol, the diffusional flux of the methanol, calculated from the following equation:

$$J_{A,s} = N_A - (N_A + N_B)\omega_{A,s} \qquad (2.11)$$

will have a positive value (direction from the interface to the bulk vapor). The slight increase in the diffusional flux with increasing amount of partially condensed vapor may be due to the high mass flux effect caused by partial condensation of vapor, as discussed in Section 4.8. The observed temperature distribution shown in Fig. 10.5 supports this reasoning.

Figure 10.7 shows mass transfer data for ternary distillation of the acetone/methanol/ethanol system on a wetted-wall column, as obtained by H. Kosuge et al. [13]. In the figure, a peculiar behavior of the mass flux of the intermediate component is observed, that is, the mass flux of methanol is not proportional to the concentration driving force, and even shows a negative finite value under conditions of zero driving force. This paradoxical behavior of the mass flux of methanol may be explained by the fact that the mass flux is the sum of the diffusional flux and the convective mass flux, as discussed in Section 2.3.

$$N_i = J_{i,s} + \rho_s v_s \omega_{i,s} = J_{i,s} + \omega_{i,s} \sum N_i \qquad (10.12)$$

Transfer of the methanol under conditions of zero driving force is caused by the convective mass flux, the second term on the right-hand side of Eq. (10.12), and the methanol is accompanied by the transfer of both acetone and ethanol.

If, on the other hand, we calculate the diffusional flux by means of Eq. (10.12), then the diffusional flux of methanol is observed to be proportional to the concentration driving force. This fact highlights an important question about the theore-

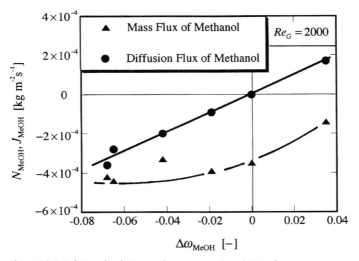

Figure 10.7 Relationship between the concentration driving force and the mass flux or diffusional flux of the intermediate component in ternary distillation of the acetone/methanol/ethanol system; data from H. Kosuge et al. [13]

tical foundation of tray efficiencies and HTUs. The apparent paradox within the classical model may be due to the fact that diffusional flux is confused with mass flux, where the former is proportional to the concentration driving force, whereas the latter is not. The peculiar behavior of tray efficiencies in Fig. 10.2 or of the numbers of transfer units in Fig. 10.3 may be due to this false assumption of the classical model.

10.2.4
Dissimilarity of Mass Transfer in Gas Absorption and Distillation

Discussions in Section 2.3 indicate that mass flux is the sum of the diffusional flux and the convective mass flux, both in gas absorption and in distillation:

Mass flux = Diffusional flux + Convective mass flux

For gas absorption, the condition of unidirectional diffusion holds and the convective mass flux can be expressed by Eq. (2.27), hence mass flux is always proportional to the concentration driving force.

$$\rho_s v_s = \frac{1}{(1-\omega_s)} J_{As} \tag{2.27}$$

In the case of multi-component distillation, however, the convective mass flux is given by Eq. (2.38):

$$\rho_s v_s = \frac{\sum_{i=1}^{i=N}(\lambda_{N,s}-\lambda_{i,s})J_{i,s} - q_G - q_w}{\sum_{i=1}^{i=N}\lambda_{i,s}\omega_{i,s}} \tag{2.38}$$

which clearly indicates that convective mass flux is not proportional to the diffusional flux due to the effects of q_G and q_w, and that the assumption of proportionality between mass flux and concentration driving force does not hold. We cannot apply tray efficiencies or HTU to predict the rate of mass transfer in multi-component distillations or in non-adiabatic binary distillations.

However, in the very special case of binary distillation under adiabatic conditions, $q_w = 0$, and negligibly small vapor-phase sensible heat flux, ($q_G \ll (\lambda_{B,s} - \lambda_{A,s})J_{A,s}$), the expression for the convective mass flux reduces to the following equation:

$$\rho_s v_s = (\lambda_{B,s} - \lambda_{A,s})J_{A,s}/(\lambda_{A,s}\omega_{A,s} + \lambda_{B,s}\omega_{B,s}) \tag{10.13}$$

Only in this special case is the convective mass flux proportional to the diffusional flux, so that the mass flux is also proportional to the concentration driving force, as is the case for gas absorption. Similar calculations as in the case of gas absorption are then possible.

Example 10.2
The following data are obtained for binary distillation of the methanol/water system on a flat-plate wetted-wall column at 1 atm under total reflux conditions.

$$N_A = -8.2 \times 10^{-3} \text{ kg m}^{-2} \text{ s}^{-1}$$
$$N_B = -2.5 \times 10^{-3} \text{ kg m}^{-2} \text{ s}^{-1}$$
$$\bar{x} = 0.75$$

Calculate the diffusional flux of the methanol under the given conditions.

Solution
From Eq. (2.23), the convective term is given by:

$$\rho_s v_s = (-8.2 \times 10^{-3}) + (-2.5 \times 10^{-3}) = -10.7 \times 10^{-3} \text{ kg m}^{-2} \text{ s}^{-1}$$

From the vapor-liquid equilibrium for the methanol/water system, the equilibrium vapor concentration for $x = 0.75$ is as follows:

$$y_s = 0.893$$

$$\omega_s = \frac{(0.893)(32)}{(0.893)(32) + (1 - 0.893)(18)} = 0.936$$

Substituting the above values into Eq. (10.10a), we have the following equation:

$$J_A = (-8.2 \times 10^{-3}) - (-10.7 \times 10^{-3})(0.936) = 1.50 \times 10^{-3} \text{ kg m}^{-2} \text{ s}^{-1}$$

This means that even in the case of negative mass flux, the diffusional flux is positive.

Example 10.3
Binary distillation of the acetone/ethanol system is carried out on a wetted-wall column of diameter 20 mm and length 600 mm at 1 atm under total reflux conditions. Calculate the liquid concentration at the top of the column, if the bottom concentration of the liquid is 0.10 and the vapor velocity is 0.5 m s^{-1}.

Solution
We will assume the following conditions:
1. Thermal insulation of the column is perfect and the vapor-phase sensible heat flux is negligibly small ($q_w = 0$, $q_G = 0$).
2. Liquid-phase resistance is negligibly small.
3. The distribution of the vapor velocity in the column is parabolic.

The diffusional fluxes can be calculated by means of the theoretical solution of the Graetz problem described in Section 5.4.

Since the present distillation is performed under total reflux conditions, the bulk liquid concentration is equal to the bulk vapor concentration.

$$x = y$$

Step 1. Vapor-liquid equilibrium of the system and physical properties of the vapor

The vapor-liquid equilibrium is the same as in Example 10.1
The physical properties of the vapor are as follows:

$$D_G = 6.66 \times 10^{-6} \text{ m}^2 \text{ s}^{-1}, \rho_G = 0.960 \text{ kg m}^{-3},$$
$$\lambda_{Acetone} = 513 \text{ kJ kg}^{-1}, \lambda_{Ethanol} = 902 \text{ kJ kg}^{-1}$$

Step 2. Preparation for the calculation

Flow rate of vapor at the bottom of the column:

$$G_1 = (0.785)(0.02)^2(0.960)(0.5) = 1.507 \times 10^{-4} \text{ kg s}^{-1}$$

Vapor concentration at the bottom of the column:

$$\omega_1 = \frac{(0.10)(58)}{(58)(0.10) + (1 - 0.10)(46)} = 0.1229$$

Flow rate of acetone at the bottom of the column:

$$G_{A,1} = (1.507 \times 10^{-4})(0.1229) = 1.85 \times 10^{-5} \text{ kg s}^{-1}$$

Flow rate of ethanol at the bottom of the column:

$$G_{B,1} = (1.507 \times 10^{-4})(1 - 0.1229) = 1.322 \times 10^{-4} \text{ kg s}^{-1}$$

From Tab. 10.1, the equilibrium vapor concentration for $x = 0.1$ is given as:

$$y_{s,1} = 0.2774,$$
$$\omega_{s,1} = (0.2774)(58)/\{(0.2774)(58) + (1 - 0.2774)(46)\} = 0.3262$$
$$\omega_{s,av} = (\omega_{s1} + \omega_{s2})/2 = (0.3262 + \omega_{s2})/2 \quad \text{(A)}$$

The logarithmic mean concentration driving force in the column:

$$\Delta\omega_{lm} = \frac{(\omega_{s2} - \omega_2) - (\omega_{s1} - \omega_1)}{\ln\{(\omega_{s2} - \omega_2)/(\omega_{s1} - \omega_1)\}} = \frac{(\omega_{s2} - \omega_2) - (0.3262 - 0.1229)}{\ln\{(\omega_{s2} - \omega_2)/(0.3262 - 0.1229)\}} \quad \text{(B)}$$

The Graetz number under the given conditions:

$$Gz_M = \frac{(3.14)(0.02)^2(0.5)}{(4)(6.66 \times 10^{-6})(0.6)} = 39.3$$

The dimensionless diffusional flux is calculated by means of Eq. (5.28):

$$\frac{J_A D_T}{\rho_G D_G \Delta\omega_{lm}} = 3.66 + \frac{(0.085)(39.3)}{1 + (0.047)(39.3)^{2/3}} = 5.82$$

The diffusional flux is given by:

$$J_A = \frac{(5.82)(0.96)(6.66 \times 10^{-6})}{(0.02)} \Delta\omega_{lm} = 1.86 \times 10^{-3} \Delta\omega_{lm} \ [\text{kg m}^{-2}\text{s}^{-1}] \quad (C)$$

Interfacial area for mass transfer:

$$S = (3.14)(0.02)(0.6) = 0.0377 \text{ m}^2$$

From Eq. (2.37), the convective mass flux is given by:

$$(\rho_s v_s)_{av} = (902 - 513)J_A/\{513\omega_{s,\,av} + 902(1 - \omega_{s,\,av})\} \quad (D)$$

The mass flux of each component is given by:

$$N_A = J_A + (\rho_s v_s)_{av}\omega_{s,\,av} \quad (E)$$

$$N_B = -J_A + (\rho_s v_s)_{av}(1 - \omega_{s,\,av}) \quad (F)$$

Flow rates of acetone and ethanol at the top of the column:

$$G_{A2} = G_{A1} + N_A S \quad (G)$$

$$G_{B2} = G_{B1} + N_B S \quad (H)$$

The concentration of vapor at the top of the column:

$$\omega_2 = G_{A1}/(G_{A2} + G_{B2}) \quad (I)$$

$$y_2 = \omega_2/(58)/\{\omega_2/(58) + (1 - \omega_2)/(46)\} \quad (J)$$

Step 3. Calculation of the vapor concentration at the top of the column

The calculation requires the logarithmic mean concentration driving forces at both ends of the column. Calculation by *a trial-and-error method* starts by tentatively assuming the initial value of the top concentration and calculations are repeated until convergence is obtained.

The first trial:
Assume $y_2 = 0.10$ ($\omega_2 = 0.1229$) as an initial value for the calculation.

$$y^* = 0.2774 \quad (\omega_{s2} = 0.3262)$$

$$\omega_{s,\,av} = 0.3262, \quad \Delta\omega_{lm} = 0.2033$$

Substituting the above values into Eqs. (C)–(J), we obtain the following value:

$$y_{2cal} = 0.1849$$

The second trial:

$$y_2 = 0.1849 \quad (\omega_2 = 0.2224), \quad y_2^* = 0.4252 \quad (\omega_{2s} = 0.4826)$$

$$\omega_{s,\,av} = (0.4826 + 0.3262)/2 = 0.4044$$

$$\Delta\omega_{lm} = \frac{\{(0.4826 - 0.2224) - (0.2033)\}}{\ln\{(0.4826 - 0.2224)/(0.2033)\}} = 0.2306$$

10.3 Simultaneous Heat and Mass Transfer Model for Packed Distillation Columns

Similar calculation gives:

$y_{2cal} = 0.2001$

The calculation is repeated until convergence of the value of y_2 is obtained. Table 10.2 shows the results of the calculation. After the fourth trial, a final convergence is obtained:

$y_2 = 0.2016$

Table 10.2 Calculation of the rate of mass transfer in binary distillation of the acetone/ethanol system on a wetted-wall column under total reflux conditions (Example 10.3)

Symbol	Unit	1st trial	2nd trial	3rd trial	4th trial
y_2	[–]	0.1000	0.1849	0.2001	0.2015
ω_2	[–]	0.1229	0.2224	0.2398	0.2414
y_{s2}	[–]	0.2774	0.4252	0.4463	0.4481
ω_{s2}	[–]	0.3262	0.4826	0.5041	0.5059
$\omega_{s,av}$	[–]	0.3262	0.4044	0.4151	0.4160
$\Delta\omega_{lm}$	[–]	0.2033	0.2306	0.2325	0.2326
J_A	[kg m^{-2} s^{-1}]	3.79E-04	4.29E-04	4.33E-04	4.33E-04
J_B	[kg m^{-2} s^{-1}]	–3.79E-04	–4.29E-04	–4.33E-04	–4.33E-04
ρv	[kg m^{-2} s^{-1}]	1.90E-04	2.34E-04	2.38E-04	2.39E-04
N_A	[kg m^{-2} s^{-1}]	4.40E-04	5.24E-04	5.32E-04	5.32E-04
N_B	[kg m^{-2} s^{-1}]	–2.51E-04	–2.90E-04	–2.93E-04	–2.94E-04
G_{A2}	[kg s^{-1}]	3.51E-05	3.83E-05	3.86E-05	3.86E-05
G_{B2}	[kg s^{-1}]	1.23E-04	1.21E-04	1.21E-04	1.21E-04
G	[kg s^{-1}]	1.58E-04	1.60E-04	1.60E-04	1.60E-04
$\omega_{2,cal}$	[–]	0.2224	0.2398	0.2414	0.2416
$y_{2,cal}$	[–]	0.1849	0.2001	0.2015	0.2016

10.3 Simultaneous Heat and Mass Transfer Model for Packed Distillation Columns

10.3.1 Wetted Area of Packings

Two major factors affecting mass transfer in packed columns are the mass fluxes of each component and the interfacial area for mass transfer. Since mass transfer in packed columns takes place on the surface of the liquid flowing over the surface of the packing, the wetted area of the packing is usually taken as the effective interfacial area, which is usually smaller than the total dry surface area of the packing due to imperfect wetting. The wetting of the surface of the packing is both a physical and chemical phenomenon, and the wetted area of the packing is affected by various factors, such as the flow rate and the physical properties of the liquid, and the shape, size, and constituent material of the packing.

K. Onda et al. [19] proposed the following correlation for the wetted area of packings commonly used in gas absorption, such as ceramic Raschig rings, Berl saddles and spheres:

$$(a_w/a_t) = 1 - \exp\{0.145\,Re_L^{0.1}\,Fr_L^{-0.05}\,We_L^{0.2}(\sigma_c/\sigma_L)^{0.75}\} \qquad (10.14)$$

where Fr is the Froude number ($Fr \equiv a_t L^2/g\rho_L^2$), Re_L is the liquid Reynolds number ($Re_L \equiv L/\mu_L a_t$), We is the Weber number ($We_L \equiv L^2/a_t\rho_L\sigma_L$), a_t is the total surface area per unit volume of the packing [m^{-1}], a_w is the wetted area [m^{-1}], L is the superficial mass velocity of the liquid [kg m^{-2} s^{-1}], σ is the surface tension of the liquid [N m^{-1}], and σ_c is the critical surface tension [N m^{-1}] characteristic for wetting of the packing material.

In contrast to gas absorption, relatively few data are available for the wetted areas of the packings used in distillation. H. Kosuge et al. [14] reported an experimental approach for determining the wetted area of 3 mm Dickson packing for various binary systems, such as the methanol/water, acetone/ethanol, acetone/methanol, and methanol/ethanol systems, over wide ranges of liquid concentrations, and reported Onda's correlation to be applicable for these systems. Figure 10.8 shows an example of their data.

The wetted area for metal-structured packings used in distillation is reported to correspond to perfect wetting under normal operating conditions of the columns [6, 18].

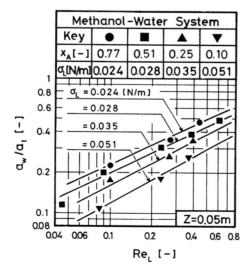

Figure 10.8 Wetted area of 3 mm Dickson packing for the methanol/water system at various liquid concentrations; data from H. Kosuge et al. [14]

10.3.2
Apparent End Effect

It is a well-known fact that laboratory-scale small packed distillation columns with the same packing but with different packed heights can sometimes yield different HTU values under the same operating conditions, and that the higher the packed height the higher the HTU. This effect is usually known as the *apparent end effect*.

Figures 10.9.a and 10.9.b show an example of an *apparent end effect*, as observed by H. Kosuge et al. [14]. They carried out binary distillations of the methanol/water and acetone/ethanol systems by using two packed columns of the same diameter

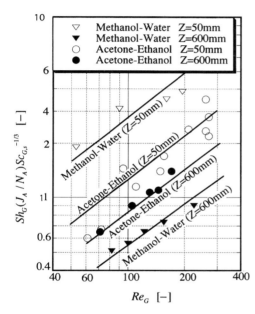

Figure 10.9a Apparent end effect in binary distillations of the methanol/water and acetone/ethanol systems

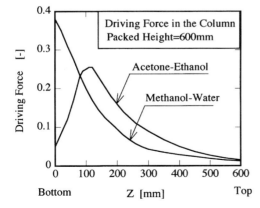

Figure 10.9b Distributions of the concentration driving forces in a 600 mm column; data from H. Kosuge et al. [14]

(22 mm inner diameter) and with the same packing (3 mm Dickson packing), but with different packed heights, 600 mm and 50 mm. Figure 10.9.a shows that the dimensionless diffusional fluxes obtained for the shorter column have consistently higher values than those obtained for the longer column, and thus an apparent end effect is observed for these systems. Figure 10.9.b shows the observed concentration driving force distributions for these systems in the 600 mm column, which were obtained by sampling the liquid at 50 mm intervals. The concentration driving force for the methanol/water system is seen to decrease exponentially, while that for the acetone/ethanol system even shows a maximum in the middle of the column. In such situations, if we calculate the average dimensionless diffusional flux from the logarithmic mean driving force, the calculated value may be quite different from the true diffusional flux, and as a result of this wrong calculation an *apparent end effect* will be observed. If, however, we use a shorter column, the variation in the concentration driving force over the length of which is less pronounced, the values obtained will be relatively consistent, and no apparent end effect will be observed. For this reason, the use of a shorter column is strongly recommended if we intend to scale-up the column on the basis of laboratory-scale data. On the other hand, variation in the concentration driving force in the case of gas absorption is usually only slight, even for a long column, and we can reliably use the logarithmic concentration driving force to determine the average dimensionless diffusional flux.

10.3.3
Correlation of the Vapor-Phase Diffusional Fluxes in Binary Distillation

Mass transfer in distillation is usually very sensitive to heat transfer through the wall of the column. If thermal insulation of the column is insufficient, mass transfer data tend not to be reliable due to partial condensation of vapor, the effect of which is highly variable in a case-by-case manner, depending on the apparatus used and the ambient temperature. Moreover, the specific surface area of a laboratory-scale column is relatively large compared to that of a commercial column; data obtained for such columns are inevitably sensitive to fluctuations in the ambient temperature, and so no general trend in the phenomena is observed. Therefore, scaling-up of packed distillation columns on the basis of data obtained for a laboratory-scale column, which usually has poor thermal insulation, will sometimes cause serious trouble.

H. Kosuge et al. [14] described a careful experimental approach to binary distillation of the acetone/ethanol system on a 22 mm diameter, 50 mm packed height column with 3 mm Dickson packing, under adiabatic conditions (without the effect of partial condensation), and proposed the following correlation for the vapor-phase local diffusional flux:

$$\{Sh_{GA}(J_{As}/N_A)\}_0 = 0.0306\, Re_G^{0.805}\, Sc_{Gs}^{1/3} \tag{10.15}$$

where the subscript 0 on the left-hand side of Eq. (10.15) denotes diffusional flux under adiabatic conditions.

10.3 Simultaneous Heat and Mass Transfer Model for Packed Distillation Columns

These authors also proposed the following empirical correlation for the effect of partial condensation of vapor by comparing Eq. (10.15) with data obtained under non-adiabatic conditions:

$$G_v \equiv \frac{\{Sh_{G,A}(J_A/N_A)\}}{\{Sh_{G,A}(J_A/N_A)\}_0} = 1 - 83.3(\rho_{Gs}v_s/\rho_{G\infty}U_\infty) \quad (10.16)$$

where v_s is the surface velocity caused by partial condensation of vapor. From Eqs. (10.15) and (10.16), the dimensionless vapor-phase local diffusional flux can be expressed as:

$$\{Sh_{GA}(J_{As}/N_A)\} = 0.0306\, Re_G^{0.805}\, Sc_{Gs}^{1/3}\, G_v \quad (10.17)$$

Figure 10.10 shows a comparison of Eq. (10.17) with the observed data, as obtained by H. Kosuge et al. [14]. The solid line in the figure represents the correlation. Data for the acetone/ethanol, methanol/ethanol, acetone/methanol, and methanol/water systems, under a wide range of operating conditions and under both adiabatic and non-adiabatic conditions, show good agreement with the proposed correlation.

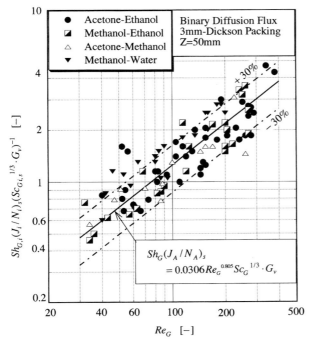

Figure 10.10 Correlation of the vapor-phase local diffusional fluxes in binary distillation on a short packed column; data from H. Kosuge et al. [14]

10.3.4
Correlation of Vapor-Phase Diffusional Fluxes in Ternary Distillation

Figure 10.11 shows the correlation of the vapor-phase local diffusional fluxes for ternary distillation of the acetone/methanol/ethanol system, as obtained by H. Kosuge et al. [15], using the same column as was used for the aforementioned binary distillation. The figure shows the dimensionless diffusional fluxes and the multi-component Schmidt numbers, which are made dimensionless by the use of Wilke's effective diffusion coefficients [8]:

$$D_{i,m} = (1 - y_{is})/(y_{j,s}/D_{G,ij} + y_{k,s}/D_{G,ik}) \tag{10.18}$$
$$(i, j, k = A, B, C)$$

The dimensionless diffusional flux:

$$Sh_{G,i}(J_i/N_i) \equiv \frac{J_{i,s} D_p}{\rho_{G,s} D_{i,m} \Delta\omega_{G,i,lm}} \tag{10.19}$$

The multi-component Schmidt number:

$$Sc_{G,i,m} \equiv \mu_{Gs}/\rho_{Gs} D_{i,m} \tag{10.20}$$

The solid line in the figure represents the binary correlation, Eq. (10.17). The observed data for the ternary distillation show good agreement with the binary correlation:

$$\{Sh_{Gi}(J_{is}/N_i)\} = 0.0306 \, Re_G^{0.805} \, Sc_{G,i,m}^{1/3} \, G_\nu \tag{10.21}$$

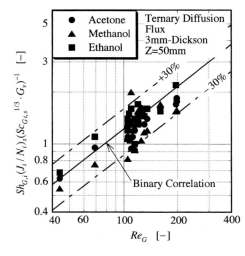

Figure 10.11 Comparison of the observed data for the vapor-phase local diffusional flux in ternary distillation of the acetone/methanol/ethanol system with the binary correlation; data from H. Kosuge et al. [15]

Similar results have been reported by S. Olano et al. [18], who found the binary and ternary distillation data for a small packed column with structured packing to be well correlated by the same correlation if Wilke's effective diffusivities are used instead of the binary diffusion coefficients. An important conclusion that emerges from these facts is that the binary correlation can be used to predict mass transfer in ternary distillation if the binary diffusion coefficients are replaced with Wilke's effective diffusion coefficients. This empirical rule is successfully applied for the cryogenic distillation of air.

10.3.5
Simulation of Separation Performance in Ternary Distillation on a Packed Column under Total Reflux Conditions

Considering the experimental facts about heat and mass transfer in binary and ternary distillations, a new approach for the prediction of separation performance in multi-component distillations on packed columns, which is based on the real heat and mass transfer phenomena in the column, can be envisaged. We refer to this approach as the *Simultaneous Heat and Mass Transfer model for distillation*. The fundamental assumptions of the model can be summarized as follows:

1. The temperature of the bulk vapor is at the dew point of the vapor, and that of the bulk liquid is at the bubble point of the liquid.
2. The vapor and the liquid are in a state of equilibrium at the interface, and the interface temperature is at the bubble point of the liquid at the interface.
3. Distillation is vapor-phase resistance controlled. In most cases, the liquid-phase resistance is negligibly small in comparison with that in the vapor phase. Even when the liquid-phase resistance is expected to have a small effect, this may be adequately estimated by the use of Higbie's penetration model.
4. The mass flux of each component and the convective mass flux are calculated according to:

$$N_i = J_i + \rho_s v_s \omega_{is} = J_i + \omega_{is} \sum_i N_i \tag{2.22}$$

$$\rho_s v_s = \frac{\sum_{i=1}^{N}(\lambda_N - \lambda_i)J_{is} - q_G - q_w}{\sum_{i=1}^{N} \lambda_i \omega_{is}} \tag{2.38}$$

For an actual calculation on a packed distillation column, the following data are needed in advance:
1. Vapor-liquid equilibrium of the system.
2. The physical properties of the system, such as the density, viscosity, and thermal conductivity of the vapor, the latent heat of vaporization of each component, and the binary diffusion coefficient for each pair of components.
3. The pressure drop and the total specific area of the packing, and correlations for the wetted area of the packing. For ordinary random packings, the wetted

area may be adequately predicted by the use of Onda's correlation [19]. Metal-structured packings, such as Mellapak, can be taken to be perfectly wetted.
4. Correlation for the vapor-phase local diffusional flux.

Calculation under conditions of total reflux is rather simple; no trial-and-error calculation of the terminal concentration is needed. Equations for operating lines are given by:

$$y_i = x_i \quad (i = A, B, C, \ldots) \tag{10.22}$$

In the following, we discuss the possibilities of a new approach based on the *Simultaneous Heat and Mass Transfer model* under total reflux conditions.

In Fig. 10.12, data for ternary distillation of the acetone/methanol/ethanol system on 22 mm diameter columns with packed heights of 50 mm and 600 mm with the same packing and under total reflux conditions, as obtained by H. Kosuge et al. [15], are compared with the simulated distillation paths derived from the *Simultaneous Heat and Mass Transfer model*. The three solid lines in the figure represent the distillation paths obtained by the proposed simulation, and the filled and unfilled circles represent the observed liquid concentrations sampled at 50 mm intervals in the 600 mm column. The short arrows represent the top and bottom concentrations obtained for the 50 mm packed height column. Good agreement between the data and the simulated distillation paths is evident.

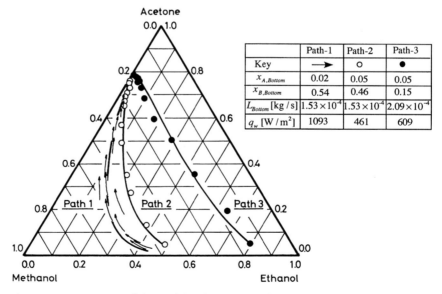

Figure 10.12 Comparison of observed data for ternary distillation of the acetone/methanol/ethanol system with the simulated distillation paths derived from the *Simultaneous Heat and Mass Transfer model*; data from H. Kosuge et al. [15]

The figure also indicates that the distillation paths cross the zero driving force line for methanol in the middle of the column, as discussed in relation to Fig. 10.3. If the classical model based on HTU were correct, the distillation path could not cross the zero driving force line, near to which the numbers of transfer units diverge infinitely. However, the data and the simulated distillation paths show quite different behavior, which suggests that there is something wrong with the classical model for distillation.

Example 10.4
Ternary distillation of the benzene/toluene/o-xylene system was carried out on a 20 mm diameter, 600 mm packed height column with 3 mm Dickson packing at 1 atm under total reflux conditions.

Calculate the concentration of each component in the distillate, if the liquid concentrations of each component in the still are as follows:

$x_A = 0.0200$, $x_B = 0.4500$, $x_C = 0.5300$

Assume that the vapor-liquid equilibrium of the system is that of an ideal system ($\gamma_A = \gamma_B = \gamma_C = 1$) and that the liquid-phase mass transfer resistance is negligibly small. Assume also that the column is under adiabatic conditions ($q_w = 0$) and that the vapor-phase sensible heat flux is negligibly small in comparison with the latent heat flux ($q_G = 0$). The apparent vapor velocity is 0.2 m s^{-1} and the pressure drop of the column is $\Delta P = 50$ mm H$_2$O m^{-1}. The specific surface area of the packing is $a_t = 2180$ m^{-1}, and the wetted area of the packing under the specified operating conditions is 50%.

Solution
Step 1. Calculation of the concentration of the bottom vapor

The vapor from the still is considered to be in equilibrium with the liquid in the still.
The total pressure in the still is given by:

$$P_{Bottom} = 101.325 + (50)(9.81)(10^{-3})(0.6) = 101.619 \text{ kPa}$$

The system is an ideal solution and the bubble point of the liquid in the still can be estimated by the use of Raoult's law, where the vapor pressure of each component is estimated by means of Antoine's equation:

Benzene (A): $\log p_A = 9.03055 - 1211.03/(T - 52.36)$

Toluene (B): $\log p_B = 9.07954 - 1344.80/(T - 53.67)$

o-Xylene (C): $\log p_B = 9.12381 - 1474.68/(T - 59.46)$

The bubble point is calculated by use of the following equations:

$$f(T) = \{(p_A x_A + p_B x_B + p_C x_C)/P\} - 1 \tag{A}$$

$$T_{i+1} = T_i - f_i/f' \tag{B}$$

The bubble point and the equilibrium vapor concentration corresponding to the bottom pressure, $P = 101.619$ kPa, are as follows:

$T_B = 396.75$ K,

$y_A = 0.0644$ ($\omega_A = 0.0528$),

$y_B = 0.6412$ ($\omega_B = 0.6195$),

$y_C = 0.2944$ ($\omega_C = 0.3277$)

The density of the vapor at the bottom of the column, $\rho_{G, Bottom}$ is given by:

$$\rho_0 = \{(3.486)(0.0644) + (4.106)(0.6412) + (4.739)(0.2944)\} = 4.252 \text{ kg m}^{-3}$$

$$\rho_{G, Bottom} = (4.252)(101.619/101.325)(273.15/396.75) = 3.205 \text{ kg m}^{-3}$$

The vapor flow rates of each component at the bottom of the column are given by:

$G_A = (0.785)(0.02)^2(3.205)(0.2)(0.0528) = 1.06 \times 10^{-5}$ kg s^{-1}

$G_B = 1.247 \times 10^{-4}$ kg s^{-1},

$G_C = 6.60 \times 10^{-5}$ kg s^{-1}

Step 2. Estimation of the physical properties of the vapor

The physical properties of the vapor are estimated at its dew point. The dew point of the vapor is calculated by application of Newton's method of convergence to obtain the temperature that will satisfy the following equation:

$$F(T) = P(y_A/p_A + y_B/p_B + y_C/p_C) - 1 \tag{C}$$

The physical properties of the vapor are also estimated at the bubble point of the liquid. The following equations are used to estimate the physical properties of the vapor:

Density of the vapor (ideal gas law):

$$\rho_{mix} = (\rho_{A0} y_A + \rho_{B0} y_B + \rho_{C0} y_C)(P/P_0)(T_0/T) \tag{D}$$

Viscosity of the pure vapor (Hirschfelder's equation):

$$\mu = 2.669 \times 10^{-6} \sqrt{MT}/\sigma^2 \Omega_\nu \tag{E}$$

Viscosity of the vapor mixture (Wilke's equation):

$$\mu_{mix} = \sum_i \mu_i y_i / \sum_j \varphi_{ij} y_j \tag{F}$$

Binary diffusion coefficient (Hirschfelder's equation):

$$D_{AB} = \frac{1.858 \times 10^{-7} T^{1.5} \sqrt{1/M_A + 1/M_B}}{(P/101325)\sigma_{AB}^2 \Omega_D} \tag{G}$$

Effective diffusion coefficient (Wilke's effective diffusion coefficient):

$$D_{im} = (1 - y_i)/(y_j/D_{ij} + y_k/D_{ik}) \tag{H}$$

Step 3. Calculation of the dimensionless diffusional flux

$$Re_G = (G_A + G_B + G_C)d_p/\mu_{G\infty} \tag{I}$$

$$Sc_i \equiv \mu_{Gs}/(\rho_{Gs} D_{i,m}) \quad (i = A, B, C) \tag{J}$$

From the assumption of adiabatic conditions for the column, $G_v = 1.0$, the dimensionless diffusional fluxes can be calculated from Eq. (10.21):

$$Sh_i(J_i/N_i) = 0.0306\, Re_G^{0.805} Sc_i^{1/3} \quad (i = A, B, C) \tag{K}$$

Step 4. Calculation of the diffusional fluxes and mass fluxes:

$$J_i = \{Sh_i(J_i/N_i)\}(\rho_{Gs} D_{i,m,s})(\omega_{G,i,s} - \omega_{G,i,\infty})/d_p \quad (i = A, B, C) \tag{L}$$

$$\rho_s v_s = \{(\lambda_{C,s} - \lambda_{A,s})J_A + (\lambda_{C,s} - \lambda_{B,s})J_B\}/(\lambda_{A,s}\omega_{A,s} + \lambda_{B,s}\omega_{B,s} + \lambda_{C,s}\omega_{C,s}) \tag{M}$$

$$N_i = J_{i,s} + (\rho_s v_s)\omega_{i,s} \quad (i = A, B, C) \tag{N}$$

Step 5. Flow rate and concentration of each component:

$$G_{i,j+1} = G_{i,j} + (\pi D_T^2/4)(a\Delta z)(N_{i,j}) \quad (i = A, B, C) \tag{O}$$

$$\omega_{i,\infty,j+1} = G_{i,j+1}/\sum G_{i,j+1} \quad (i = A, B, C) \tag{P}$$

$$y_{i,j+1} = (\omega_{i,j+1}/M_i)/\sum \omega_{i,j+1}/M_i \quad (i = A, B, C) \tag{Q}$$

$$P_j = P_{j-1} - \Delta P_{packing}\Delta Z \tag{R}$$

The results of the calculation using the above equations are shown in Fig. 10.13 and Tabs. 10.3a–e. Table 10.3e shows that the vapor flow rate at the top of the column is 2% lower than that at the bottom. This may be due to the fact that the convective mass fluxes given by Eq. (M) are negative in all regions of the column.

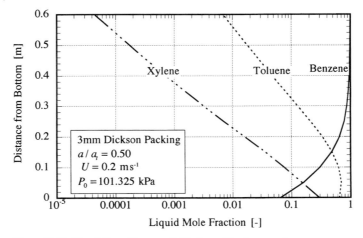

Figure 10.13 Simulation of ternary distillation of the benzene/toluene/o-xylene system on a 600 mm packed column under total reflux conditions by the *Simultaneous Heat and Mass Transfer model*

Table 10.3a Temperature and pressure distributions and vapor concentration in the column (Example 10.4)

z [m]	P [kPa]	T_D [K]	y_A [-]	y_B [-]	y_C [-]
0.00	101.619	396.75	0.064448	0.641178	0.294374
0.05	101.595	388.98	0.158307	0.691699	0.149995
0.10	101.570	382.13	0.302619	0.622488	0.074893
0.15	101.546	375.36	0.480625	0.483775	0.035600
0.20	101.521	368.90	0.652419	0.331548	0.016033
0.25	101.497	363.56	0.783784	0.209134	0.007082
0.30	101.472	359.76	0.869193	0.127611	0.003196
0.35	101.448	357.28	0.920974	0.077533	0.001493
0.40	101.423	355.72	0.951952	0.047330	0.000718
0.45	101.399	354.75	0.970598	0.029050	0.000352
0.50	101.374	354.15	0.981923	0.017902	0.000175
0.55	101.350	353.78	0.988847	0.011065	0.000088
0.60	101.325	353.55	0.993103	0.006853	0.000044

Table 10.3b Vapor-liquid equilibrium (Example 10.4)

z [m]	T_B [K]	y_A^* [–]	y_B^* [–]	y_C^* [–]
0.00	387.48	0.165995	0.711005	0.123000
0.05	379.77	0.335213	0.616013	0.048849
0.10	372.72	0.531385	0.449539	0.019216
0.15	366.30	0.706490	0.286294	0.007282
0.20	361.28	0.830300	0.166958	0.002729
0.25	357.96	0.904510	0.094381	0.001064
0.30	355.99	0.945632	0.053907	0.000445
0.35	354.86	0.968404	0.031525	0.000199
0.40	354.20	0.981236	0.018820	0.000093
0.45	353.80	0.988503	0.011396	0.000045
0.50	353.56	0.992932	0.006967	0.000022
0.55	353.41	0.995571	0.004285	0.000011
0.60	353.32	0.997331	0.002646	0.000006

Table 10.3c Calculation of dimensionless diffusional flux (Example 10.4)

z [m]	Re_G [–]	Sc_A [–]	Sc_B [–]	Sc_C [–]	$Sh_A(J_A/N_A)$ [–]	$Sh_B(J_B/N_B)$ [–]	$Sh_C(J_C/N_C)$ [–]
0.00	272.9	0.558	0.614	0.689	2.304	2.377	2.471
0.05	235.6	0.630	0.647	0.771	2.131	2.149	2.279
0.10	226.9	0.681	0.683	0.818	2.121	2.124	2.255
0.15	225.7	0.718	0.718	0.848	2.150	2.149	2.272
0.20	221.5	0.743	0.742	0.866	2.142	2.141	2.254
0.25	212.4	0.758	0.757	0.876	2.084	2.084	2.188
0.30	202.7	0.766	0.765	0.881	2.014	2.014	2.111
0.35	195.2	0.773	0.770	0.884	1.961	1.957	2.050
0.40	190.1	0.779	0.772	0.886	1.924	1.918	2.008
0.45	186.8	0.771	0.774	0.886	1.890	1.893	1.980
0.50	184.7	0.766	0.775	0.887	1.869	1.876	1.963
0.55	183.3	0.752	0.775	0.887	1.847	1.866	1.951
0.60	182.5	0.771	0.776	0.887	1.855	1.859	1.944

Table 10.3d Calculation of diffusional fluxes, convective mass fluxes, and mass fluxes (Example 10.4)

J_A [kg m^{-2} s^{-1}]	J_B [kg m^{-2} s^{-1}]	J_C [kg m^{-2} s^{-1}]	$(\rho v)_s$ [kg m^{-2} s^{-1}]	N_A [kg m^{-2} s^{-1}]	N_B [kg m^{-2} s^{-1}]	N_C [kg m^{-2} s^{-1}]
9.88E-04	1.00E-03	−1.79E-03	−1.51E-04	9.67E-04	8.96E-04	−1.81E-03
1.60E-03	−4.68E-04	−9.79E-04	−1.26E-04	1.56E-03	−5.49E-04	−9.86E-04
2.07E-03	−1.46E-03	−5.36E-04	−1.18E-04	2.01E-03	−1.51E-03	−5.39E-04
2.09E-03	−1.78E-03	−2.78E-04	−1.05E-04	2.02E-03	−1.82E-03	−2.79E-04
1.67E-03	−1.52E-03	−1.32E-04	−7.88E-05	1.60E-03	−1.54E-03	−1.32E-04
1.12E-03	−1.05E-03	−5.85E-05	−5.14E-05	1.07E-03	−1.06E-03	−5.86E-05
6.89E-04	−6.60E-04	−2.61E-05	−3.13E-05	6.59E-04	−6.62E-04	−2.61E-05
4.15E-04	−4.03E-04	−1.20E-05	−1.87E-05	3.97E-04	−4.04E-04	−1.20E-05
2.51E-04	−2.46E-04	−5.68E-06	−1.12E-05	2.40E-04	−2.46E-04	−5.68E-06
1.54E-04	−1.51E-04	−2.76E-06	−6.90E-06	1.47E-04	−1.51E-04	−2.76E-06
9.48E-05	−9.26E-05	−1.36E-06	−4.25E-06	9.06E-05	−9.26E-05	−1.36E-06
5.91E-05	−5.71E-05	−6.81E-07	−2.67E-06	5.64E-05	−5.72E-05	−6.81E-07
3.59E-05	−3.53E-05	−3.42E-07	−1.60E-06	3.43E-05	−3.54E-05	−3.42E-07

Table 10.3e Distribution of mass velocities and mass fractions of each component of the vapor in the column (Example 10.4)

z [m]	G_A [kg s^{-1}]	G_B [kg s^{-1}]	G_C [kg s^{-1}]	G_{total} [kg s^{-1}]	ω_A [−]	ω_B [−]	ω_C [−]
0.00	1.062E-05	1.247E-04	6.596E-05	2.01E-04	0.0528	0.6195	0.3277
0.00	2.72E-05	1.40E-04	3.50E-05	2.02E-04	0.1344	0.6926	0.1731
0.05	5.38E-05	1.31E-04	1.81E-05	2.03E-04	0.2658	0.6448	0.0894
0.10	8.82E-05	1.05E-04	8.88E-06	2.02E-04	0.4371	0.5189	0.0440
0.15	1.23E-04	7.36E-05	4.10E-06	2.01E-04	0.6124	0.3671	0.0205
0.20	1.50E-04	4.73E-05	1.85E-06	1.99E-04	0.7536	0.2372	0.0093
0.25	1.69E-04	2.92E-05	8.43E-07	1.99E-04	0.8488	0.1470	0.0042
0.30	1.80E-04	1.79E-05	3.96E-07	1.98E-04	0.9078	0.0902	0.0020
0.35	1.87E-04	1.09E-05	1.91E-07	1.98E-04	0.9437	0.0553	0.0010
0.40	1.91E-04	6.74E-06	9.41E-08	1.98E-04	0.9654	0.0341	0.0005
0.45	1.93E-04	4.16E-06	4.68E-08	1.98E-04	0.9787	0.0210	0.0002
0.50	1.95E-04	2.57E-06	2.35E-08	1.97E-04	0.9869	0.0130	0.0001
0.55	1.96E-04	1.59E-06	1.18E-08	1.97E-04	0.9919	0.0081	0.0001
0.60	1.96E-04	9.89E-07	5.99E-09	1.97E-04	0.9950	0.0050	0.0000

Example 10.5

Further to Example 10.4, calculate the concentration of each component and the flow rate of the vapor at the top of the column under non-adiabatic conditions, if $q_w = 0.02$ kW m^{-2}, $q_w = 0.04$ kW m^{-2}, $q_w = 0.06$ kW m^{-2}, $q_w = 0.08$ kW m^{-2}.

Assume that the operating conditions are the same as in Example 10.4.

Solution

The calculation is almost the same as in the case of Example 10.4, except that the convective mass flux is calculated from the following equation rather than Eq. (M):

$$\rho_s v_s = \frac{(\lambda_{C,s} - \lambda_{A,s})J_A + (\lambda_{C,s} - \lambda_{B,s})J_B - q_w}{(\lambda_{A,s}\omega_{A,s} + \lambda_{B,s}\omega_{B,s} + \lambda_{C,s}\omega_{C,s})}$$

Table 10.4 shows the results of the calculation. This example shows that the partial condensation of vapor due to heat loss from the wall has a considerable effect on the concentration and the flow rate of the vapor at the top of the column. This suggests that special care must be taken with regard to the effects of heat loss if we intend to scale-up a plant on the basis of data obtained for a laboratory-scale small column, the thermal insulation of which is usually imperfect.

Table 10.4 Effect of wall heat flux on separation performance in ternary distillation on packed columns (Example 10.5)

q_w [kW m^{-2}]	$y_{A,Top}$ [–]	$y_{B,Top}$ [–]	$y_{C,Top}$ [–]	G_{Top}/G_{bottom} [–]
0 (adiabatic)	0.993103	0.006853	0.000044	0.980
0.02	0.994024	0.005941	0.000035	0.935
0.04	0.994875	0.005098	0.000027	0.886
0.06	0.995681	0.004299	0.000020	0.840
0.08	0.996419	0.003566	0.000015	0.791

10.4 Calculation of Ternary Distillations on Packed Columns under Finite Reflux Ratio

10.4.1 Material Balance for the Column

Figure 10.14 shows the material balance for a packed distillation column. The total material balance for the column and that for each component can be expressed by the following equations:

$$F = D + B \tag{10.23}$$

$$f_i = d_i + b_i \tag{10.24}$$

From the definition of the reflux ratio, the vapor and liquid flow rates at the top of the column can be expressed by the following equations:

$$V_{Top} = (R+1)D \tag{10.25}$$

$$L_{Top} = RD \tag{10.26}$$

where B, D, and F are the mass flow rates of the so-called bottoms, the distillate, and the feed [kg m^{-2} s^{-1}], respectively, and b_i, d_i, and f_i are the mass flow rates of component i in the bottoms, the distillate, and the feed. L_{Top} and V_{Top} are the mass flow rates of the liquid and the vapor at the top of the column [kg m^{-2} s^{-1}], and R is the reflux ratio.

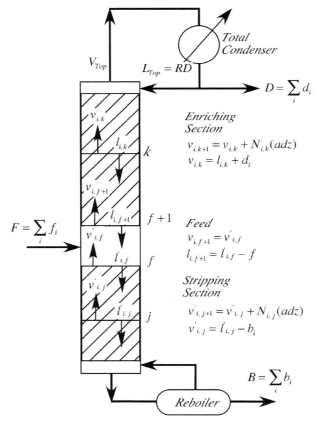

Figure 10.14 Material balance for a distillation column

10.4 Calculation of Ternary Distillations on Packed Columns under Finite Reflux Ratio

Expressions for the material balance in the *Enriching Section* can be written as follows:

$$v_{i,k} = l_{i,k} + d_i \quad (10.27)$$

$$v_{i,k+1} = v_{i,k} + N_{i,k}\, a dZ \quad (10.28)$$

$$l_{i,k+1} = l_{i,k} + N_{i,k}\, a dZ \quad (10.29)$$

If we assume that the feed is supplied as the liquid at its bubble point, the material balance for the feed section can be expressed by the following equations:

$$v_{i,f} = v'_{i,f} \quad (10.30)$$

$$l_{i,f} = l'_{i,f} - f_i \quad (10.31)$$

Expressions for the material balance in the *Stripping Section* can be written as follows:

$$v'_{i,j} = l'_{i,j} - b_i \quad (10.32)$$

$$v'_{i,j+1} = v'_{i,j} + N_{i,j}\, a dZ \quad (10.33)$$

$$l'_{i,j+1} = l'_{i,j} + N_{i,j}\, a dZ \quad (10.34)$$

The reboiler or the still is assumed to be an equilibrium stage:

$$y_{i,0} = y_i^*(x_{i,\,Btm}) \quad (10.35)$$

The bottom concentration necessary for calculation of the equilibrium vapor concentration from the reboiler is given by the following equation:

$$b_i = f_i - d_i$$

$$x_{i,B} = \frac{(b_i/M_i)}{\sum_i (b_i/M_i)} \quad (10.36)$$

where a is the interfacial area for mass transfer [m^2/m^3], $l_{i,k}$ and $l'_{i,j}$ are the mass flow rates of component i in the liquid in the enriching section and the stripping section, respectively [kg m^{-2} s^{-1}], and $v_{i,k}$ and $v'_{i,j}$ are the mass flow rates of component i in the vapor in the enriching section and the stripping section, respectively [kg m^{-2} s^{-1}]. $N_{i,j}$ and $N_{i,k}$ are the mass fluxes of component i in section j of the stripping section and in the section k of the enriching section [kg m^{-2} s^{-1}]. The mass flux of each component is calculated by means of the following equations, if correlations for the vapor-phase diffusional fluxes and interfacial area are known:

$$N_i = J_i + \rho_s v_s \omega_{is} \tag{2.22}$$

$$\rho_s v_s = \frac{\sum_{i=1}^{N}(\lambda_N - \lambda_i)J_{is} - q_G - q_w}{\sum_{i=1}^{N}\lambda_i \omega_{is}} \tag{2.38}$$

10.4.2
Convergence of Terminal Composition

Calculation under conditions of a finite reflux ratio is rather more complicated than that under total reflux conditions and requires a trial-and-error approach as in the case of an equilibrium stage model. However, if suitable initial values for the composition of the distillate are known *a priori* by some means, usually by considering the results of a total reflux, then the calculation proceeds in a similar way as that under total reflux conditions, that is, from the reboiler to the top of the column, and the calculation is repeated until convergence in the composition of the distillate is obtained.

Convergence of the terminal concentrations is easily obtained by applying the so-called θ-method of convergence devised by W. N. Lyster et al. [16], which is based on the total material balance and the material balance for each component at both ends of the column. The condition is that the top and bottom mass flow rates must satisfy the constraints of Eqs. (10.23) and (10.24). At the end of each iteration, if the top and bottom mass flow rates, d_i and b_i, of each component do not satisfy Eq. (10.24), they should be corrected in order to satisfy the material balance by introducing a correction factor θ, as defined in the following equation:

$$d_{i,co} = \frac{f_i}{1 + \theta(b_i/d_i)_{Cal}} \tag{10.37}$$

From Eqs. (10.23) and (10.37), we obtain the following equation for θ:

$$g(\theta) = \sum_i \frac{(f_i/F)}{1 + \theta(b_i/d_i)_{Cal}} - (D/F) \tag{10.38}$$

The condition for satisfying the total material balance, Eq. (10.23), is then reduced to obtaining the solution for Eq. (10.38) in terms of θ:

$$g(\theta) = 0$$

Substituting the numerical value of θ into Eq. (10.37), the corrected value for the distillate, $d_{i,co}$, is obtained and the next iteration is started. Calculation is repeated until the condition that $d_i = d_{i,Cal}$ ($i = 1, 2, 3, \ldots N$) is obtained ($\theta = 1.000$). Figure 10.15 shows a flow chart for the calculation.

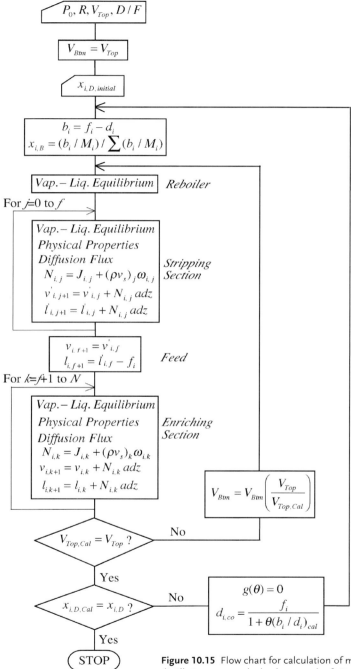

Figure 10.15 Flow chart for calculation of multi-component distillation on a packed column under finite reflux ratio by means of the *Simultaneous Heat and Mass Transfer model*

10 Mass Transfer in Distillation

Example 10.6

Ternary distillation of a 60:30:10 mixture of benzene, toluene, and o-xylene was carried out on a 20 mm diameter, 600 mm packed height column with 3 mm Dickson packing (a_t = 2180 m^{-1}) at 1 atm. Calculate the distribution of each component in the column under the following conditions:

Packed height of enriching section = 0.3 m
Packed height of stripping section = 0.3 m
Reflux ratio = 2.0
Vapor flow rate at the top of the column = 0.600 kg m^{-2} s^{-1}
D/F = 0.5
ΔP = 50 mm H$_2$O m^{-1}
a/a_t = 0.50

Solution

We assume an ideal system and adiabatic conditions for the column. The physical properties of the system are the same as those in Examples 10.4 and 10.5. The diffusional fluxes are estimated by means of Eq. (10.21).

Step 1. Preparation:

$$D = 0.6000/(2+1) = 0.2000 \text{ kg m}^{-2}\text{s}^{-1}$$

$$F = (0.2000)/(0.5) = 0.4000 \text{ kg m}^{-2}\text{s}^{-1}$$

$$f_A = \frac{(0.4000)(0.6000)(78.11)}{(0.6000)(78.11) + (0.3000)(92.14) + (0.1000)(106.17)} = 0.2202 \text{ kg m}^{-2}\text{s}^{-1}$$

$$f_B = 0.1299 \text{ kg m}^{-2}\text{s}^{-1}$$

$$f_C = 0.0499 \text{ kg m}^{-2}\text{s}^{-1}$$

Step 2. First trial:

As initial values for the distillate, we assume:

$$x_A = 0.9600, \quad x_B = 0.0390, \quad x_C = 0.0010$$

$$d_A = \frac{(0.2000)(0.9600)(78.11)}{(0.9600)(78.11) + (0.0390)(92.14) + (0.0010)(106.17)} = 0.1906 \text{ kg m}^{-2}\text{s}^{-1}$$

$$d_B = 0.0091 \text{ kg m}^{-2}\text{s}^{-1}$$

$$d_C = 0.0003 \text{ kg m}^{-2}\text{s}^{-1}$$

$$b_A = 0.2202 - 0.1906 = 0.0296 \text{ kg m}^{-2}\text{s}^{-1}$$

$$b_B = 0.1299 - 0.0091 = 0.1208 \text{ kg m}^{-2}\text{s}^{-1}$$

$$b_C = 0.0499 - 0.0003 = 0.0496 \text{ kg m}^{-2}\text{s}^{-1}$$

10.4 Calculation of Ternary Distillations on Packed Columns under Finite Reflux Ratio

As an initial value for the bottom mass flow rate, we assume:

$V_{Bottom} = V_{Top} = 0.6000$ kg m^{-2} s^{-1}

The bottom-to-top calculation proceeds in a similar way as in Example 10.4 by using the material balance, Eqs. (10.27)–(10.34), together with the correlation for diffusional flux. After the trial, we have $V_{Top,Cal} = 0.5861$ kg m^{-2} s^{-1}, which is different from the initial value. We then repeat a similar calculation using the corrected value for $V_{Bottom,Co}$, as given by the following equation:

$V_{Bottom,Co} = V_{Bottom}(0.6000/0.5861) = 0.6142$ kg m^{-2} s^{-1}

At the end of the calculation, the following values are obtained:

$(b_A/d_A)_{Cal} = 0.1494$

$(b_B/d_B)_{Cal} = 71.09$

$(b_C/d_C)_{Cal} = 2564$

Substituting these values into Eq. (10.38), we obtain the following equation:

$$\frac{(0.2202)/(0.4000)}{1 + 0.1494\theta} + \frac{(0.1299)/(0.4000)}{1 + 71.09\theta} + \frac{(0.0499)/(0.4000)}{1 + 2564\theta} - 0.5 = 0$$

Solution of the above equation in terms of θ is as follows:

$\theta = 0.7651$

Substituting the above value into Eq. (10.37), we have the following values for the second trial:

$d_A = (0.2202)/\{1 + (0.7651)(0.1494)\} = 0.1976$ kg m^{-2} s^{-1}

$d_B = (0.1299)/\{1 + (0.7651)(71.09)\} = 0.0023$ kg m^{-2} s^{-1}

$d_C = (0.0499)/\{1 + (0.7651)(2564)\} = 0.0000$ kg m^{-2} s^{-1}

The corrected top concentrations for the second trial are as follows:

$x_{A,D} = 0.9899$, $x_{B,D} = 0.0100$, $x_{C,D} = 0.0001$

Step 3. The second trial to final convergence

Similar calculations are repeated until full convergence is obtained for the terminal concentrations ($\theta = 1$).

2nd trial:	$x_{A,D} = 0.9899$,	$x_{B,D} = 0.0100$,	$x_{C,D} = 0.0001$,	$\theta = 1.1448$
3rd trial:	$x_{A,D} = 0.9754$,	$x_{B,D} = 0.0245$,	$x_{C,D} = 0.0001$,	$\theta = 0.9461$
4th trial:	$x_{A,D} = 0.9815$,	$x_{B,D} = 0.0183$,	$x_{C,D} = 0.0002$,	$\theta = 1.0253$
5th trial:	$x_{A,D} = 0.9787$,	$x_{B,D} = 0.0211$,	$x_{C,D} = 0.0002$,	$\theta = 0.9881$
6th trial:	$x_{A,D} = 0.9800$,	$x_{B,D} = 0.0199$,	$x_{C,D} = 0.0001$,	$\theta = 1.0053$

7th trial: $x_{A,D} = 0.9794$, $x_{B,D} = 0.0204$, $x_{C,D} = 0.0002$, $\theta = 0.9975$
8th trial: $x_{A,D} = 0.9796$, $x_{B,D} = 0.0202$, $x_{C,D} = 0.0002$, $\theta = 1.0008$

Table 10.5 shows the results for final convergence. Figure 10.16 shows the distribution of each component in the column.

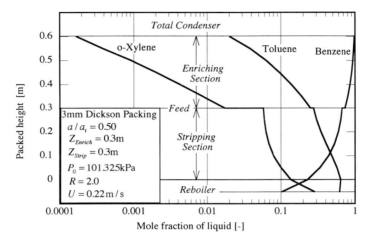

Figure 10.16 Simulation of ternary distillation of the benzene/toluene/o-xylene system on a 600 mm packed column under finite reflux conditions by means of the *Simultaneous Heat and Mass Transfer model* (Example 10.6)

Table 10.5a Temperature and pressure distributions and vapor concentration in the column (Example 10.6)

z [m]	P [kPa]	y_A [-]	y_B [-]	y_C [-]	x_A [-]	x_B [-]	x_C [-]
Still	101.619				0.1041	0.6141	0.2818
0.00	101.619	0.2529	0.6378	0.1093	0.2281	0.6365	0.1354
0.05	101.595	0.3909	0.5456	0.0635	0.3343	0.5659	0.0998
0.10	101.570	0.5252	0.4369	0.0380	0.4386	0.4819	0.0795
0.15	101.546	0.6371	0.3385	0.0244	0.5262	0.4053	0.0686
0.20	101.521	0.7208	0.2619	0.0173	0.5920	0.3453	0.0626
0.25	101.497	0.7789	0.2076	0.0135	0.6379	0.3027	0.0594
0.30	101.472	0.8173	0.1712	0.0115	0.6683	0.2740	0.0577
0.30	101.472	0.8173	0.1712	0.0115	0.7334	0.2492	0.0174
0.35	101.448	0.8613	0.1328	0.0058	0.8008	0.1905	0.0087
0.40	101.423	0.8983	0.0988	0.0029	0.8571	0.1386	0.0043
0.45	101.399	0.9279	0.0707	0.0014	0.9018	0.0962	0.0020
0.50	101.374	0.9505	0.0488	0.0007	0.9359	0.0631	0.0009
0.55	101.350	0.9505	0.0488	0.0007	0.9612	0.0384	0.0004
0.60	101.325	0.9796	0.0203	0.00017			

10.4 Calculation of Ternary Distillations on Packed Columns under Finite Reflux Ratio

Table 10.5b Vapor-liquid equilibrium (Example 10.6)

z [m]	P [kPa]	T_B [K]	y_A^* [–]	y_B^* [–]	y_C^* [–]
Still	101.619	385.14	0.2529	0.6378	0.1093
0.00	101.619	376.59	0.4442	0.5162	0.0397
0.05	101.595	372.00	0.5754	0.3996	0.0250
0.10	101.570	368.20	0.6799	0.3027	0.0174
0.15	101.546	365.38	0.7535	0.2329	0.0136
0.20	101.521	363.43	0.8020	0.1865	0.0115
0.25	101.497	362.14	0.8327	0.1568	0.0104
0.30	101.472				
0.30	101.472	359.26	0.8798	0.1175	0.0027
0.35	101.448	357.57	0.9138	0.0849	0.0013
0.40	101.423	356.25	0.9403	0.0591	0.0006
0.45	101.399	355.25	0.9601	0.0397	0.0003
0.50	101.374	354.51	0.9745	0.0254	0.0001
0.55	101.350	353.98	0.9848	0.0152	0.0000
0.60	101.325				

Table 10.5c Calculation of dimensionless diffusional flux (Example 10.6)

z [m]	Re_G [–]	Sc_A [–]	Sc_B [–]	Sc_C [–]	$Sh_A(J_A/N_A)$ [–]	$Sh_B(J_B/N_B)$ [–]	$Sh_C(J_C/N_C)$ [–]
Still							
0.00	195.7	0.632	0.641	0.764	1.837	1.845	1.956
0.05	198.3	0.684	0.687	0.816	1.906	1.909	2.021
0.10	199.4	0.708	0.707	0.834	1.937	1.936	2.045
0.15	199.8	0.724	0.722	0.846	1.954	1.953	2.058
0.20	200.0	0.735	0.731	0.853	1.965	1.963	2.065
0.25	200.1	0.742	0.738	0.857	1.972	1.969	2.070
0.30							
0.30	200.1	0.753	0.751	0.871	1.983	1.981	2.081
0.35	200.6	0.760	0.759	0.877	1.993	1.992	2.090
0.40	201.0	0.765	0.764	0.880	2.000	1.999	2.096
0.45	201.2	0.769	0.768	0.883	2.005	2.004	2.099
0.50	201.3	0.772	0.771	0.885	2.008	2.008	2.102
0.55	201.1	0.773	0.773	0.886	2.008	2.008	2.101
0.60							

Table 10.5d Diffusional flux, convective mass flux, and mass flux (Example 10.6)

J_A [kg m^{-2} s^{-1}]	J_B [kg m^{-2} s^{-1}]	J_C [kg m^{-2} s^{-1}]	(ρv) [kg m^{-2} s^{-1}]	N_A [kg m^{-2} s^{-1}]	N_B [kg m^{-2} s^{-1}]	N_C [kg m^{-2} s^{-1}]
1.50E-03	−8.32E-04	−5.81E-04	−9.72E-05	1.46E-03	−8.85E-04	−5.85E-04
1.51E-03	−1.14E-03	−3.34E-04	−8.32E-05	1.47E-03	−1.17E-03	−3.37E-04
1.30E-03	−1.10E-03	−1.82E-04	−6.57E-05	1.26E-03	−1.12E-03	−1.83E-04
1.00E-03	−8.94E-04	−9.75E-05	−4.81E-05	9.66E-04	−9.06E-04	−9.83E-05
7.06E-04	−6.51E-04	−5.21E-05	−3.31E-05	6.81E-04	−6.58E-04	−5.26E-05
4.73E-04	−4.43E-04	−2.81E-05	−2.18E-05	4.55E-04	−4.47E-04	−2.84E-05
5.60E-04	−4.70E-04	−8.21E-05	−2.80E-05	5.36E-04	−4.74E-04	−8.22E-05
4.72E-04	−4.25E-04	−4.29E-05	−2.25E-05	4.52E-04	−4.27E-04	−4.29E-05
3.78E-04	−3.55E-04	−2.18E-05	−1.75E-05	3.62E-04	−3.56E-04	−2.18E-05
2.91E-04	−2.79E-04	−1.09E-05	−1.32E-05	2.78E-04	−2.80E-04	−1.09E-05
2.17E-04	−2.11E-04	−5.36E-06	−9.76E-06	2.08E-04	−2.12E-04	−5.36E-06
1.58E-04	−1.55E-04	−2.61E-06	−7.05E-06	1.51E-04	−1.55E-04	−2.61E-06

Table 10.5e Distribution of the mass velocities of each component of the vapor and liquid in the column (Example 10.6)

z [m]	v_A [kg m^{-2} s^{-1}]	v_B [kg m^{-2} s^{-1}]	v_C [kg m^{-2} s^{-1}]	l_A [kg m^{-2} s^{-1}]	l_B [kg m^{-2} s^{-1}]	l_C [kg m^{-2} s^{-1}]
0.00	0.1348	0.4011	0.0792	0.1598	0.5262	0.1290
0.05	0.2143	0.3528	0.0473	0.2394	0.4780	0.0971
0.10	0.2945	0.2889	0.0289	0.3195	0.4141	0.0788
0.15	0.3633	0.2277	0.0189	0.3883	0.3528	0.0688
0.20	0.4159	0.1783	0.0136	0.4409	0.3034	0.0634
0.25	0.4530	0.1424	0.0107	0.4780	0.2676	0.0605
0.30	0.4778	0.1180	0.0092	0.5028	0.2432	0.0590
0.30	0.4778	0.1180	0.0092	0.2826	0.1133	0.0091
0.35	0.5070	0.0922	0.0047	0.3118	0.0875	0.0046
0.40	0.5316	0.0689	0.0023	0.3364	0.0642	0.0023
0.45	0.5513	0.0496	0.0011	0.3562	0.0448	0.0011
0.50	0.5665	0.0343	0.0006	0.3713	0.0296	0.0005
0.55	0.5778	0.0228	0.0003	0.3826	0.0180	0.0002
0.60	0.5861	0.0143	0.0001	0.3909	0.0095	0.0001

10.5
Cryogenic Distillation of Air on Packed Columns

10.5.1
Air Separation Plant

A conventional air separation plant is composed of a *double-column distillation* set-up, which comprises a *high-pressure column* and a *low-pressure column*, an *argon column*, and an expansion turbine for cold supply. The *high-pressure column* is operated at 5 atm and serves as a heat source for the *low-pressure column* through a reboiler-condenser and also provides the oxygen-enriched feed supply to the *low-pressure column*. The *low-pressure column* is operated at atmospheric pressure and serves as a separator for product-grade oxygen and nitrogen. The *argon column* is operated at atmospheric pressure, and serves to purify the argon-rich feed, a side-cut from the *low-pressure column*, to give product-grade argon.

A characteristic of the air separation plant is that both the raw material and the heat source to the reboiler and the condenser are provided by compressed air. Therefore, reducing operating costs depends solely on being able to reduce the outlet pressure of the compressor, which means reducing the pressure drop in the *low-pressure column* and the *argon column*, both of which are operated at atmospheric pressure. Until the late 1980s, tray towers were commonly used in air separation plants, but the recent development of highly efficient column packings has led to a reevaluation of the merits of packed columns in the air separation industry. Indeed, the use of such packed columns as the aforementioned *low-pressure column* and *argon column* is becoming increasingly popular.

10.5.2
Mass and Diffusional Fluxes in Cryogenic Distillation

N. Egoshi et al. [6] reported an experimental approach to mass transfer in the binary distillation of nitrogen/oxygen and argon/oxygen systems for a wide range of operating conditions and liquid concentrations under total reflux conditions using a 208 mm diameter, 1035 mm packed height distillation column with Mellapak 500Y structured packing. Figure 10.17 shows the effects of the bottom concentration of oxygen on the normalized mass flux, $N_A/\Delta\omega_A$, and the normalized diffusional flux, $J_A/\Delta\omega_A$, respectively. The figure indicates that $N_A/\Delta\omega_A$ decreases linearly with oxygen concentration, whereas $J_A/\Delta\omega_A$ is almost uniform, and the difference between the two lines may be attributed to the effect of convective mass fluxes. It was also reported that Eq. (2.37) holds for the convective mass fluxes in cryogenic binary distillation.

Figure 10.18 shows data for binary cryogenic distillations of the nitrogen/oxygen and argon/oxygen systems on the same packed column, obtained by N. Egoshi et al. [6]. The solid line in the figure represents the correlation for the vapor-phase local diffusional flux:

$$Sh_{G,sA}(J_{G,sA}/N_A) = 0.0075 \, Re_G^{'0.98} Sc_{G,s}^{1/3} \tag{10.37}$$

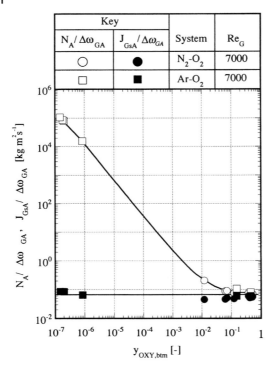

Figure 10.17 Mass and diffusional flux in binary cryogenic distillation on a packed column; data from N. Egoshi et al. [6]

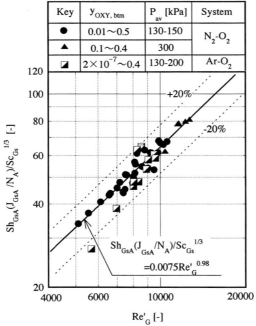

Figure 10.18 Correlation of vapor-phase binary cryogenic distillations on a packed distillation column; data from N. Egoshi et al. [6]

where Re'_G is the vapor-phase Reynolds number defined by the relative velocity of the vapor with respect to the surface velocity of the liquid:

$$U_{Geff} \equiv \frac{U_G}{(\varepsilon - a_t \delta_L)\sin\theta} + u_{Ls} \tag{10.38}$$

$$Re'_G \equiv \rho_G U_{Geff} d_h / \mu_G \tag{10.39}$$

The fact that the data for wide ranges of liquid concentrations and operating pressures are well correlated by the above correlation is worthy of note.

10.5.3
Simulation of Separation Performance of a Pilot-Plant-Scale Air Separation Plant

N. Egoshi et al. [6] reported an experimental approach to the separation performance of a pilot-plant-scale air separation plant with packed columns. Table 10.6 shows the specifications of the plant. A parallel simulation approach to the separation performance of the plant was carried out by applying the *Simultaneous Heat and Mass Transfer model*, where the ternary diffusional fluxes were estimated by use of the binary correlation, Eq. (10.37), replacing the binary diffusion coefficients with Wilke's effective diffusion coefficients.

Table 10.6 Specifications of a pilot-plant-scale air separation plant

Distillation Column	Column Internals	Column Diameter (mm)	Number of Trays or Packed Height
High-pressure column	sieve tray	710	unspecified
Low-pressure column	MellaPak 500Y	500	18.0 m
Argon column	MellaPak 500Y	380	11.1 m

In Fig. 10.19, the observed data are compared with the results of simulation for the *low-pressure column*. The simulated distributions for the nitrogen, oxygen, and argon concentrations show excellent agreement with the data. The two peaks for the argon concentration represent the effect of side-cut feed to the *argon column* and of its return from the *argon column*. It was also reported that the simulation for the *argon column* likewise showed excellent agreement with the observed data [7].

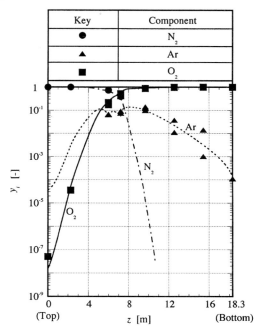

Figure 10.19 Comparison of the observed data for the low-pressure column in a pilot-plant-scale air separation plant with the results of simulation by the *Simultaneous Heat and Mass Transfer model*; data from N. Egoshi et al. [6]

10.6
Industrial Separation of Oxygen-18 by Super Cryogenic Distillation

10.6.1
Oxygen-18 as Raw Material for PET Diagnostics

Oxygen-18 is one of the three stable isotopes of oxygen, the others being oxygen-16 and oxygen-17. Natural oxygen is therefore a six-component system of diatomic molecules of the three isotopes. Table 10.7 shows the composition of natural oxygen, which indicates that $^{16}O^{18}O$ (4000 ppm) is the principal source of oxygen-18 compounds, rather than $^{18}O_2$, which is only a trace component (4.2 ppm). Since

Table 10.7 Isotopic composition of natural oxygen

Oxygen isotope	Molecular weight	Composition
$^{16}O_2$	32	0.99519
$^{16}O^{17}O$	33	0.00074
$^{16}O^{18}O$	34	0.00407
$^{17}O_2$	34	0.1 ppm
$^{17}O^{18}O$	35	1.5 ppm
$^{18}O_2$	36	4.2 ppm

all three isotopes of oxygen are stable, the concentration of oxygen-18 in chemical compounds is essentially the same as that in natural oxygen, 0.2 atom%.

Laboratory-scale plants for the production of oxygen-18 water ($H_2^{18}O$) through the distillation of water or nitrogen monoxide were developed in the U.S. and Israel for unspecified purposes in the late 1960s, but commercial demand for the product has been quite limited until very recently. The advent of *Positron Emission Tomography* (PET) in the medical sector has led to a drastic increase in the commercial need for oxygen-18 water. Nowadays, medical imaging with PET has attained worldwide recognition as one of the safest, most precise, and most reliable diagnostic methods for detecting cancers in the human body. Oxygen-18 water supplied by the manufacturers is converted into $H^{18}F$ in dedicated medical cyclotrons, which is chemically converted to 2-[^{18}F]-fluoro-2-deoxy-D-glucose at PET clinics and finally administered to patients for diagnostic purposes. Since the method is quick and reliable, even painless to the patients, PET diagnostics is now widely used in the U.S., Europe, and Japan, and as a result the commercial demand for its raw material, oxygen-18 water, $H_2^{18}O$, continues to increase rapidly. However, the conventional processes are too involved for successful scale-up and their product is of rather low purity, hence they cannot meet this urgent and socially important need. The development of a new process for the stable and reliable supply of high-purity oxygen-18 water is becoming an important social issue.

10.6.2
A New Process for Direct Separation of Oxygen-18 from Natural Oxygen

A new process for the direct separation of oxygen-18 from natural oxygen, according to:

Natural oxygen → $^{16}O^{18}O$ → $^{18}O_2$ → $H_2^{18}O$

has recently been developed by Taiyo-Nissan Corporation, Japan (formerly Nippon Sanso Corporation). The new process has the following merits over conventional processes.
1. The process is simple, so it is easy to scale-up and to construct a large capacity plant.
2. Simple process flow enables lower hold-up and shorter start-up period.
3. Low energy consumption is expected for the process because of the low heat of vaporization of the oxygen, which is as low as one-sixth of that of water, and so the energy consumption of the plant will be accordingly low.
4. The direct separation enables high purity of the product.

In spite of these technical merits, the following difficulties were expected in the process development:
1. Extremely fine separation of a six-component system of extremely low relative volatility, where the difference between the normal boiling points of the most volatile and the least volatile components is less than 1 K.

2. Low-temperature operation at around 90 K and an extremely large influx of heat from the environment are expected, necessitating due consideration of the effect of heat transfer on mass transfer.
3. High degree of enrichment of the feed, almost 500-fold.
4. Development of an isotope-exchange reactor, which means finding a suitable catalyst and the optimum temperature for the following reaction:

$$^{16}O^{18}O \rightarrow \tfrac{1}{4}{}^{16}O_2 + \tfrac{1}{2}{}^{16}O^{18}O + \tfrac{1}{4}{}^{18}O_2$$

The difficulties of this separation process could be overcome by applying the *Simultaneous Heat and Mass Transfer model* described in the previous sections of this chapter in conjunction with a pilot-plant study on the diffusional fluxes in the system. An optimum reactor design was made possible through a laboratory study.

Figure 10.20 shows a schematic flow diagram of the new process for the separation of oxygen-18 from natural oxygen. The set-up is composed of five major units, an oxygen purifier for supply of ultra-pure grade oxygen, a preliminary distillation unit for enrichment of $^{16}O^{18}O$, an isotope exchange reactor for converting $^{16}O^{18}O$ to $^{18}O_2$, a main distillation unit for the enrichment of $^{18}O_2$, and a hydrogenation reactor for production of $H_2^{18}O$, the raw material for PET diagnostics. The oxygen purifier is a conventional *double-column* type tray tower, wherein commercial grade oxygen from a neighboring air separation plant is purified to ultra-pure grade oxygen with impurities at less than 1 ppb. The preliminary distillation unit and the main distillation unit are subdivided into 13 packed columns. The large diameter distillation columns are filled with a structured packing, but the small diameter columns are filled with a specially manufactured random packing. Each column is equipped with a condenser, a reboiler, and a liquid reservoir, which enables easy restart of the plant in case of emergency. The cold supply to the plant is

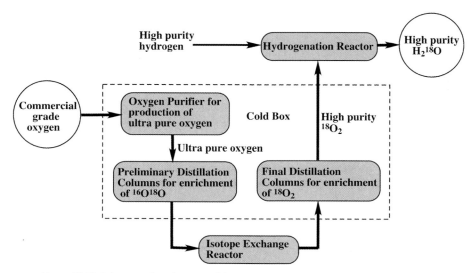

Figure 10.20 Schematic flow diagram of the new oxygen-18 separation process

provided by circulating commercial grade nitrogen with a nitrogen compressor and a expansion turbine. The distillation columns are housed in a 7 m × 3.5 m cross-section and 70 m high cold box with pearlite insulation. The waste oxygen and nitrogen are recycled to the air separation plant, making the plant zero-emission in nature.

10.6.3 Construction and Operation of the Plant

The project started in early 2001 with financial support from the Japan Science and Technology Agency. After three-and-a-half years of basic design, pilot-plant study, construction and test operation, the plant started commercial operation. During construction of the plant, a dynamic simulator capable of simulating the behavior of the plant, such as the pressure, temperature, and flow rates of the vapor and the liquid, and the vapor and liquid concentrations of all six component dioxygen isotopic species, in both an unsteady state and a steady state, was developed by use of the *Simultaneous Heat and Mass Transfer model*. Figure 10.21 shows a comparison of the observed distribution of the concentrations of oxygen isotopes in the plant with the simulated distribution. The observed isotope concentrations show remarkably good agreement with the results of the simulation. The concentration distribution during the start-up period (unsteady operation) also showed good agreement with the simulated distribution. The most difficult separation in the world has thus been accomplished by application of the *Simultaneous Heat and Mass Transfer model*, which we may refer to as the *Super Cryogenic Distillation* of oxygen isotopes.

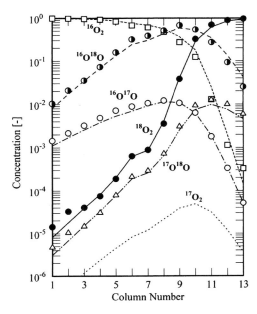

Figure 10.21 Comparison of observed data for the separation of oxygen isotopes in the oxygen-18 separation plant with the simulated distribution based on the *Simultaneous Heat and Mass Transfer model*

Figure 10.22 Picture of the oxygen-18 separation plant (Courtesy of Taiyo-Nissan Corporation)

Table 10.8 Specifications of the oxygen-18 separation plant

Cold box	7 m × 3.5 m × 70 m
Thermal insulation of cold box	Pearlite insulation
Energy supply	Nitrogen circulation with compressor and expansion turbine
Oxygen purifier	Double-column tray tower
Preliminary and main distillation unit	13 packed columns with structured packing and random packing
Operating pressure	Atmospheric pressure
Operating temperature	90 K
Capacity of the plant	100 kg per annum of high purity $H_2^{18}O$ for medical use (99 atom%)
	plus 100 kg per annum of low purity $H_2^{18}O$ for industrial use (10 atom%)
Start-up period	180 days

Figure 10.22 shows a picture of the oxygen-18 separation plant. Table 10.7 lists the specifications of the plant. The plant is now in steady-state operation and is producing 100 kg of high purity oxygen-18 water ($H_2^{18}O$) for medical use (99 atom%) annually, almost one-third of the world's annual commercial needs for PET diagnostics, as well as a further 100 kg of low purity $H_2^{18}O$ for industrial use (10 atom%). The energy consumption of the plant is as anticipated at the beginning of the project; indeed, the operating cost of the plant may be the lowest in the world, whereas the quality of the product is the highest.

References

1 A. I. Ch. E. Research Committee, "*Tray Efficiencies in Distillation Columns*", University of Delaware (1958).

2 K. Asano, H. Kihara, T. Kambe, S. Hayashida, and H. Kawakami, "Industrial Separation of Oxygen-18 from Natural Oxygen by Super Cryogenic Distillation", *Proceedings of the 7th World Congress of Chemical Engineering*, paper No. O85-004, Glasgow, U.K. (2005).

3 M. W. Bidulph and M. M. Dribika, "Distillation Efficiencies on a Large Sieve Plate with Small-Diameter Holes", *AIChE Journal*, **32**, [8], 1383–1388 (1986).

4 T. H. Chilton and A. P. Colburn, "Distillation and Absorption in Packed Columns: A Convenient Design and Correlation Method", *Industrial and Engineering Chemistry*, **27**, [3], 255–260 (1935).

5 N. Egoshi, H. Kawakami, and K. Asano, "Mass Transfer in Ternary Distillation of Nitrogen/Argon/Oxygen System in Wetted-Wall Column", *Journal of Chemical Engineering of Japan*, **32**, [4], 409–416 (1999).

6 N. Egoshi, H. Kawakami, and K. Asano, "Mass Transfer in Binary Distillation of Nitrogen/Oxygen and Argon/Oxygen Systems by Packed Column with Structured Packings", *Journal of Chemical Engineering of Japan*, **33**, [2], 245–252 (2000).

7 N. Egoshi, H. Kawakami, and K. Asano, "Heat and Mass Transfer Model Approach to Prediction of Separation Performance of Cryogenic Air Separation Plant by Packed Columns with Structured Packing", *Journal of Chemical Engineering of Japan*, **34**, [1], 22–29 (2001).

8 D. F. Fairbanks and C. R. Wilke, "Diffusion Coefficients in Multicomponent Gas Mixtures", *Industrial and Engineering Chemistry*, **42**, [3], 471–475 (1950).

9 G. G. Haselden and R. M. Thorogood, "Point Efficiency in the Distillation of the Oxygen/Nitrogen/Argon System", *Transactions of Institutions of Chemical Engineers*, **42**, 81–100 (1964).

10 A. Ito and K. Asano, "Simultaneous Heat and Mass Transfer in Binary Distillation", *Kagaku Kogaku Ronbunshu*, **6**, [4], 352–358 (1980); *International Chemical Engineering*, **22**, [2], 309–318 (1982).

11 A. Ito and K. Asano, "Thermal Effects in Non-Adiabatic Binary Distillation; Effects of Partial Condensation of Mixed Vapors on the Rates of Heat and Mass Transfer and Prediction of H. T. U.", *Chemical Engineering Science*, **37**, [4], 1007–1014 (1982)

12 H. Kosuge and K. Asano, "Mass and Heat Transfer in Ternary Distillation of Methanol/Ethanol/Water Systems by a Wetted-Wall Column", *Journal of Chemical Engineering of Japan*, **15**, [4], 268–273 (1982).

13 H. Kosuge, T. Johkoh, and K. Asano, "Experimental Studies of Diffusion Fluxes of Ternary Distillation of Acetone/Methanol/Ethanol Systems by a Wetted-Wall Column", *Chemical Engineering Communications*, **34**, [1], 111–122 (1985).

14 H. Kosuge, J. Matsudaira, K. Aoki, and K. Asano, "Experimental Approach to Mass Transfer in Binary Packed Column Distillation", *Journal of Chemical Engineering of Japan*, **23**, [5], 593–599 (1990).

15. H. Kosuge, J. Matsudaira, and K. Asano, "Ternary Mass Transfer in Packed Distillation Column", *Journal of Chemical Engineering of Japan*, **24**, [4], 455–460 (1991).
16. W. N. Lyster, S. L. Sullivan, Jr., D. S. Billingsley, and C. D. Holland, "Figure Distillation This New Way: Part 1. New Convergence Method Will Handle Many Cases", *Petroleum Refiner*, **38**, [5], 221 (1959).
17. E. V. Murphree, "Rectifying Column Calculations: With Particular Reference to N Component Mixtures", *Industrial and Engineering Chem*istry, **17**, 747–750 (1925).
18. S. Olano, Jr., S. Nagura, H. Kosuge, and K. Asano, "Mass Transfer in Binary and Ternary Distillation by a Packed Column with Structured Packing", *Journal of Chemical Engineering of Japan*, **28**, [6], 750–757 (1995).
19. K. Onda, H. Takeuchi, and Y. Koyama, "Effect of Packing Materials on the Wetted Surface Area", *Kagaku Kogaku*, **31**, [2], 126–133 (1967).
20. A. Vogelpohl, "Murphree Efficiencies in Multi-component Systems", Distillation, 1979, *Vol. 1* (I. Chem. E. Symposium Series No. 56), 2.1, 25–31 (1979).

Appendix

A. Governing Equations in Cylindrical and Spherical Coordinates

Table A1   Cylindrical Coordinates
Table A2   Spherical Coordinates

B. Phase Equilibrium Data

Figure A1   Henry's Constants of Common Gases in Water
Figure A2   Vapor Pressures of Some Hydrocarbons
Figure A3   Humidity Chart

C. Physical Properties

Table A3   Summary of Physical Properties of Gases
Table A4   Property Data
Figure A4   Viscosities of Common Gases
Figure A5   Diffusivities of Gases in Air

A. Governing Equations in Cylindrical and Spherical Coordinates

Table A1 Governing equations in cylindrical coordinates for a flow of constant physical properties without chemical reaction[a].

Continuity equation:

$$\frac{1}{r}\frac{\partial}{\partial r}(rv_r) + \frac{1}{r}\frac{\partial}{\partial \theta}(v_\theta) + \frac{\partial v_z}{\partial z} = 0 \qquad \text{(A-1)}$$

Equation of motion:

$$\frac{\partial v_r}{\partial t} + v_r\frac{\partial v_r}{\partial r} + \frac{v_\theta}{r}\frac{\partial v_r}{\partial \theta} - \frac{v_\theta^2}{r} + v_z\frac{\partial v_r}{\partial z}$$

$$= -\frac{1}{\rho}\frac{\partial P}{\partial r} + \frac{\mu}{\rho}\left\{\frac{\partial}{\partial r}\left(\frac{1}{r}\frac{\partial}{\partial r}(rv_r)\right) + \frac{1}{r^2}\frac{\partial^2 v_r}{\partial \theta^2} - \frac{2}{r^2}\frac{\partial v_\theta}{\partial \theta} + \frac{\partial^2 v_r}{\partial z^2}\right\} + g_r \qquad \text{(A-2a)}$$

$$\frac{\partial v_\theta}{\partial t} + v_r\frac{\partial v_\theta}{\partial r} + \frac{v_\theta}{r}\frac{\partial v_\theta}{\partial \theta} + \frac{v_r v_\theta}{r} + v_z\frac{\partial v_\theta}{\partial z}$$

$$= -\frac{1}{\rho}\frac{\partial P}{r\partial \theta} + \frac{\mu}{\rho}\left\{\frac{\partial}{\partial r}\left(\frac{1}{r}\frac{\partial}{\partial r}(rv_\theta)\right) + \frac{1}{r^2}\frac{\partial^2 v_\theta}{\partial \theta^2} + \frac{2}{r^2}\frac{\partial v_r}{\partial \theta} + \frac{\partial^2 v_\theta}{\partial z^2}\right\} + g_\theta \qquad \text{(A-2b)}$$

$$\frac{\partial v_z}{\partial t} + v_r\frac{\partial v_z}{\partial r} + \frac{v_\theta}{r}\frac{\partial v_z}{\partial \theta} + v_z\frac{\partial v_z}{\partial z}$$

$$= -\frac{1}{\rho}\frac{\partial P}{\partial z} + \frac{\mu}{\rho}\left\{\frac{1}{r}\frac{\partial}{\partial r}\left(r\frac{\partial v_z}{\partial r}\right) + \frac{1}{r^2}\frac{\partial^2 v_z}{\partial \theta^2} + \frac{\partial^2 v_z}{\partial z^2}\right\} + g_z \qquad \text{(A-2c)}$$

Energy equation (without viscous dissipation):

$$\frac{\partial T}{\partial t} + v_r\frac{\partial T}{\partial r} + \frac{v_\theta}{r}\frac{\partial T}{\partial \theta} + v_z\frac{\partial T}{\partial z} = \left(\frac{\kappa}{\rho c_p}\right)\left\{\frac{1}{r}\frac{\partial}{\partial r}\left(r\frac{\partial T}{\partial r}\right) + \frac{1}{r^2}\frac{\partial^2 T}{\partial \theta^2} + \frac{\partial^2 T}{\partial z^2}\right\} \qquad \text{(A-3)}$$

Diffusion equation (binary system without chemical reaction):

$$\frac{\partial \omega_A}{\partial t} + v_r\frac{\partial \omega_A}{\partial r} + \frac{v_\theta}{r}\frac{\partial \omega_A}{\partial \theta} + v_z\frac{\partial \omega_A}{\partial z} = D\left\{\frac{1}{r}\frac{\partial}{\partial r}\left(r\frac{\partial \omega_A}{\partial r}\right) + \frac{1}{r^2}\frac{\partial^2 \omega_A}{\partial \theta^2} + \frac{\partial^2 \omega_A}{\partial z^2}\right\} \qquad \text{(A-4)}$$

[a] R. B. Bird, W. E. Stewart, and E. N. Lightfoot, „Transport Phenomena", pp. 83–91, 318–319, 559, Wiley (1960).

Table A2 Governing equations in spherical coordinates (r, θ, ϕ) for a flow of constant physical properties without chemical reaction[a].

Continuity equation:

$$\frac{1}{r^2}\frac{\partial}{\partial r}(r^2 v_r) + \frac{1}{r\sin\theta}\frac{\partial}{\partial \theta}(v_\theta \sin\theta) + \frac{1}{r\sin\theta}\frac{\partial}{\partial \phi}(v_\phi) = 0 \qquad (A\text{-}5)$$

Equation of motion:

$$\frac{\partial v_r}{\partial t} + v_r \frac{\partial v_r}{\partial r} + \frac{v_\theta}{r}\frac{\partial v_r}{\partial \theta} + \frac{v_\phi}{r\sin\theta}\frac{\partial v_r}{\partial \phi} - \frac{v_\theta^2 + v_\phi^2}{r} = -\frac{1}{\rho}\frac{\partial P}{\partial r}$$

$$+ \frac{\mu}{\rho}\left\{\nabla^2 v_r - \frac{2}{r^2} v_r - \frac{2}{r^2}\frac{\partial v_\theta}{\partial \theta} - \frac{2}{r^2} v_\theta \cot\theta - \frac{2}{r^2 \sin\theta}\frac{\partial v_\phi}{\partial \phi}\right\} + g_r \qquad (A\text{-}6a)$$

$$\frac{\partial v_\theta}{\partial t} + v_r \frac{\partial v_\theta}{\partial r} + \frac{v_\theta}{r}\frac{\partial v_\theta}{\partial \theta} + \frac{v_\phi}{r\sin\theta}\frac{\partial v_\theta}{\partial \phi} + \frac{v_r v_\theta}{r} - \frac{v_\phi^2 \cot\theta}{r} = -\frac{1}{\rho}\frac{\partial P}{r\partial \theta}$$

$$+ \frac{\mu}{\rho}\left\{\nabla^2 v_\theta + \frac{2}{r^2}\frac{\partial v_r}{\partial \theta} - \frac{v_\theta}{r^2 \sin^2\theta} - \frac{2\cos\theta}{r^2 \sin^2\theta}\frac{\partial v_\phi}{\partial \phi}\right\} + g_\theta \qquad (A\text{-}6b)$$

$$\frac{\partial v_\phi}{\partial t} + v_r \frac{\partial v_\phi}{\partial r} + \frac{v_\theta}{r}\frac{\partial v_\phi}{\partial \theta} + \frac{v_\phi}{r\sin\theta}\frac{\partial v_\phi}{\partial \phi} + \frac{v_r v_\phi}{r} - \frac{v_\theta v_\phi \cot\theta}{r} = -\frac{1}{\rho}\frac{1}{r\sin\theta}\frac{\partial P}{\partial \phi}$$

$$+ \frac{\mu}{\rho}\left\{\nabla^2 v_\phi - \frac{v_\phi}{r^2 \sin^2\theta} + \frac{2}{r^2 \sin^2\theta}\frac{\partial v_r}{\partial \phi} + \frac{2\cos\theta}{r^2 \sin^2\theta}\frac{\partial v_\theta}{\partial \phi}\right\} + g_\phi \qquad (A\text{-}6c)$$

Energy equation (without viscous dissipation):

$$\frac{\partial T}{\partial t} + v_r \frac{\partial T}{\partial r} + \frac{v_\theta}{r}\frac{\partial T}{\partial \theta} + \frac{v_\phi}{r\sin\theta}\frac{\partial T}{\partial \phi}$$

$$= \left(\frac{\kappa}{\rho c_p}\right)\left\{\frac{1}{r^2}\frac{\partial}{\partial r}\left(r^2 \frac{\partial T}{\partial r}\right) + \frac{1}{r^2 \sin\theta}\frac{\partial}{\partial \theta}\left(\sin\theta \frac{\partial T}{\partial \theta}\right) + \frac{1}{r^2 \sin^2\theta}\frac{\partial^2 T}{\partial \phi^2}\right\} \qquad (A\text{-}7)$$

Diffusion equation (binary system without chemical reaction):

$$\frac{\partial \omega_A}{\partial t} + v_r \frac{\partial \omega_A}{\partial r} + \frac{v_\theta}{r}\frac{\partial \omega_A}{\partial \theta} + \frac{v_\phi}{r\sin\theta}\frac{\partial \omega_A}{\partial \phi}$$

$$= D\left\{\frac{1}{r^2}\frac{\partial}{\partial r}\left(r^2 \frac{\partial \omega_A}{\partial r}\right) + \frac{1}{r^2 \sin\theta}\frac{\partial}{\partial \theta}\left(\sin\theta \frac{\partial \omega_A}{\partial \theta}\right) + \frac{1}{r^2 \sin^2\theta}\frac{\partial^2 \omega_A}{\partial \phi^2}\right\} \qquad (A\text{-}8)$$

$$\nabla^2 = \frac{1}{r^2}\frac{\partial}{\partial r}\left(r^2 \frac{\partial}{\partial r}\right) + \frac{1}{r^2 \sin\theta}\frac{\partial}{\partial \theta}\left(\sin\theta \frac{\partial}{\partial \theta}\right) + \frac{1}{r^2 \sin^2\theta}\left(\frac{\partial^2}{\partial \phi^2}\right) \qquad (A\text{-}9)$$

[a] R. B. Bird, W. E. Stewart, and E. N. Lightfoot, „Transport Phenomena", pp. 83–91, 318–319, 559, Wiley (1960).

B. Phase Equilibrium Data

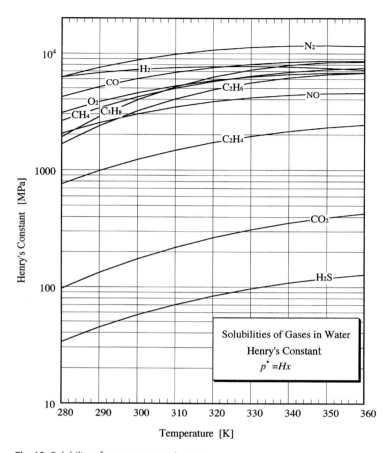

Fig. A1 Solubility of common gases in water.

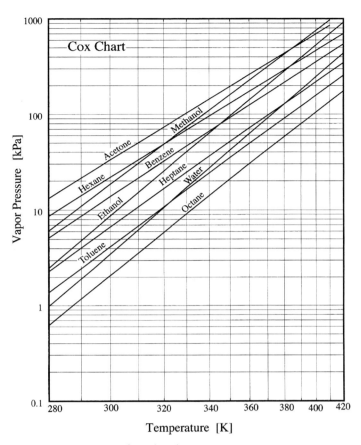

Fig. A2 Vapor pressures of pure liquids.

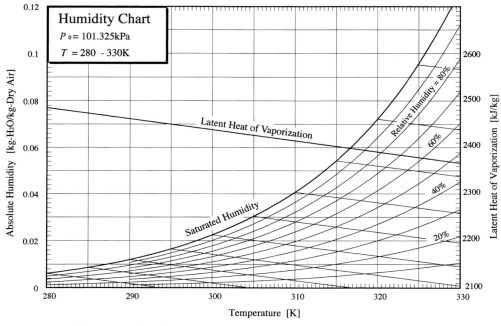

Fig. A3 Humidity chart.

C. Physical Properties

Table A3 Summary of estimation of physical properties of gases.

	Property	Unit	Estimation equation
Pure component	Density	kg m^{-3}	$\rho = (P/RT)(M/1000)$
	Molar heat at constant pressure	J mol^{-1} K^{-1}	$C_p = a + bT + cT^2 + dT^3$
	Viscosity	Pa s	Hirschfelder's equation $$\mu = \frac{2.669 \times 10^{-6}\sqrt{MT}}{\sigma^2 \Omega_v(T^*)}$$ $$\Omega_v(T^*) = \frac{1.16145}{(T^*)^{0.14874}} + \frac{0.52487}{\exp(0.77320T^*)} + \frac{2.16178}{\exp(2.43787T^*)}$$ $T^* = T/(\varepsilon/k)$
	Thermal conductivity	W m^{-1} K^{-1}	Eucken's equation $(\kappa/\mu)(M/1000) = C_v + 2.25R = C_p + 1.25R$
Mixture	Density	kg m^{-3}	$\rho_{mix} = \sum_i^N y_i \rho_i$
	Molar heat at constant pressure	J mol^{-1} K^{-1}	$C_{p\,mix} = \sum_{i=1}^N y_i C_{pi}$
	Viscosity	Pa s	Wilke's equation $$\mu_{mix} = \sum_i \frac{y_i \mu_i}{\sum_j y_j \phi_{ij}}$$ $$\phi_{ij} = \frac{\left\{1 + (\mu_j/\mu_i)^{1/2}(M_j/M_i)^{1/4}\right\}^2}{2.8284(1 + M_i/M_j)^{1/2}}$$
	Thermal conductivity	W m^{-1} K^{-1}	Wilke's equation $$\kappa_{mix} = \sum_i \frac{y_i \kappa_i}{\sum_j y_j \phi_{ij}}$$ $$\phi_{ij} = \frac{\left\{1 + (\mu_j/\mu_i)^{1/2}(M_j/M_i)^{1/4}\right\}^2}{2.8284(1 + M_i/M_j)^{1/2}}$$
	Diffusivity	m^2 s^{-1}	Hirschfelder's equation $$D_{AB} = \frac{1.858 \times 10^{-3} T^{2/3}\sqrt{1/M_A + 1/M_B}}{(P/101325)\sigma_{AB}^2 \Omega_D(T_D^*)}$$ $$\Omega_D(T_D^*) = \frac{1.06036}{(T_D^*)^{0.15610}} + \frac{0.19300}{\exp(0.47635T_D^*)} + \frac{1.03587}{\exp(1.52996T_D^*)} + \frac{1.76474}{\exp(3.89411T_D^*)}$$ $\sigma_{AB} = (\sigma_A + \sigma_B)/2$ $T_D^* = T/\sqrt{(\varepsilon_A/k)(\varepsilon_B/k)}$

Notation:

C_p = molar heat at constant pressure [J mol^{-1} K^{-1}]
D_{AB} = binary diffusion coefficient [m^2 s^{-1}]
M = molecular weight [kg kmol^{-1}]
P = pressure [Pa]
R = Gas constant (8.31433 J mol^{-1} K^{-1})
T = temperature [K]
κ = thermal conductivity [W m^{-1} K^{-1}]
μ = viscosity [Pa s]
ρ = density [kg m^{-3}]
σ = collision diameter [10^{-10} m]
ε/k = molecular constant [K]

Table A.4 Property data.

Material	Chemical formula	Molecular weight	Molecular constant		Density		Normal boiling point	Latent heat of vaporization
		M	σ	ε/k	$\rho_{G,NTP}$	$\rho_{L,298\,K}$	T_b	$L_{v,298\,K}$
		[kg kmol^{-1}]	[10^{-10} m]	[K]	[kg m^{-3}]	[kg m^{-3}]	[K]	[kJ mol^{-1}]
Methane	CH_4	16.04	3.80	144	0.716		111.7	
Ethane	C_2H_6	30.07	4.42	230	1.342		184.5	
Propane	C_3H_8	44.10	5.06	254	1.986		231.1	
Butane	C_4H_{10}	58.12	5.00	410	2.593		272.7	
Pentane	C_5H_{12}	72.15	5.78	344.1		626	309.2	26.43
Hexane	C_6H_{14}	86.18	5.91	413		659	341.9	31.55
Heptane	C_7H_{16}	100.21	–	–		684	371.6	36.55
Octane	C_8H_{18}	114.23	7.41	333		703	398.8	41.49
Ethylene	C_2H_4	28.05	4.23	205	1.251		169.4	
Propylene	C_3H_6	42.08	4.67	303	1.877		225.5	
Benzene	C_6H_6	78.11	5.27	440		885	353.3	33.85
Toluene	C_7H_8	92.14	5.93	377		867	383.8	37.99
Methanol	CH_3OH	32.04	3.67	452		791	337.8	37.50
Ethanol	C_2H_5OH	46.07	4.37	415		789	351.5	42.30
Formaldehyde	HCOH	30.03			1.340		254.0	
Acetone	CH_3COCH_3	58.08	5.05	417	2.591	790	329.4	30.80
Air	Air	28.97	3.62	97	1.293		79.2	
Hydrogen	H_2	2.02	2.93	37	0.090		20.4	
Water	H_2O	18.02	2.65	356	0.804	999.7	373.2	44.01
Nitrogen	N_2	28.01	3.68	91.5	1.250		77.4	
Nitrogen monoxide	NO	30.01	3.49	119	1.339		121.4	
Oxygen	O_2	32.00	3.54	88	1.428		90.2	
Sulfur dioxide	SO_2	64.06	4.11	335.4	2.858		263.0	
Carbon monoxide	CO	28.01	3.59	110	1.250		81.7	
Carbon dioxide	CO_2	44.01	4.00	190	1.964		194.7	

Table A.4 Property data (continued).

Chemical formula	Molar heat at constant pressure $C_p = a + bT + cT^2 + dT^3$				Antoine's constant $\log p = A - B/(T + C)$			
	a	b	c	d	A	B	C	Range
	[J mol^{-1} K^{-1}]	[J mol^{-1} K^{-2}]	[J mol^{-1} K^{-3}]	[J mol^{-1} K^{-4}]	[–]	[K]	[K]	[K]
CH_4	19.89	5.02E-02	1.27E-05	–1.10E-08	8.82051	405.42	–5.37	91–120
C_2H_6	6.90	1.73E-01	–6.41E-05	7.29E-09	8.95942	663.70	–16.68	130–198
C_3H_8	–4.04	3.05E-01	–1.57E-04	3.17E-08	8.92888	803.81	–26.16	164–247
C_4H_{10}	3.96	3.71E-01	–1.83E-04	3.50E-08	8.93386	935.86	–34.42	195–292
C_5H_{12}	6.77	4.54E-01	–2.25E-04	4.23E-08	9.00122	1075.78	–39.95	223–330
C_6H_{14}	6.94	5.52E-01	–2.87E-04	5.77E-08	8.99514	1168.72	–48.94	247–365
C_7H_{16}	–5.15	6.76E-01	–3.65E-04	7.66E-08	9.01875	1264.37	–56.51	271–396
C_8H_{18}	–6.10	7.71E-01	–4.20E-04	8.86E-08	9.03430	1349.82	–63.77	292–425
C_2H_4	3.95	1.56E-01	–8.34E-05	1.77E-08	8.87246	585.00	–18.15	119–181
C_3H_6	3.15	2.38E-01	–1.22E-04	2.46E-08	8.94450	785.00	–26.15	161–241
C_6H_6	–36.22	4.85E-01	–3.14E-04	7.76E-08	9.03055	1211.03	–52.36	280–377
C_7H_8	–34.39	5.59E-01	–3.45E-04	8.04E-08	9.07954	1344.80	–53.67	279–409
CH_3OH	19.05	9.15E-02	–1.22E-05	–8.04E-09	10.00350	1473.11	–43.15	253–413
C_2H_5OH	19.89	2.10E-01	–1.04E-04	2.01E-08	10.16980	1554.30	–50.50	270–369
HCOH	22.81	4.08E-02	7.13E-06	–8.70E-09	9.28100	957.24	–30.15	185–271
CH_3COCH_3	6.80	2.79E-01	–1.56E-04	3.48E-08	9.14937	1161.00	–49.15	257–376
Air	28.11	1.97E-03	4.80E-06	–1.97E-09				
H_2	29.11	–1.92E-03	4.00E-06	–8.70E-10	8.04578	71.62	3.19	13–521
H_2O	32.24	1.92E-03	1.06E-05	–3.51E-09	10.09170	1668.21	–45.15	333–423
N_2	28.90	–1.57E-03	8.08E-06	–2.87E-09	9.56802	333.17	–2.28	54–80
NO	29.34	–9.40E-04	9.75E-06	–4.19E-09	12.32360	874.13	–0.05	95–419
O_2	25.48	1.52E-02	–7.16E-06	1.31E-09	8.81634	319.01	–6.45	62–449
SO_2	25.78	5.79E-02	–3.81E-05	8.61E-09	9.40718	999.90	–35.96	195–279
CO	28.16	1.68E-03	5.37E-06	–2.22E-09	8.36510	230.27	–13.15	63–108
CO_2	22.26	5.98E-02	–3.50E-05	7.47E-09	11.76670	1284.07	–4.72	154–204

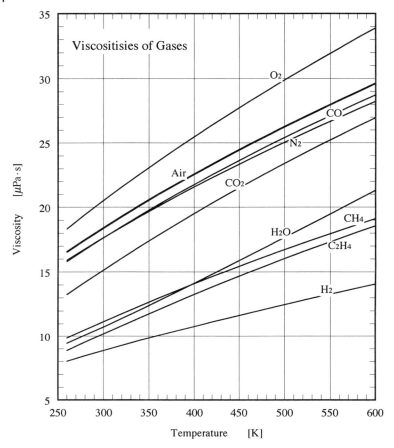

Fig. A4 Viscosities of gases.

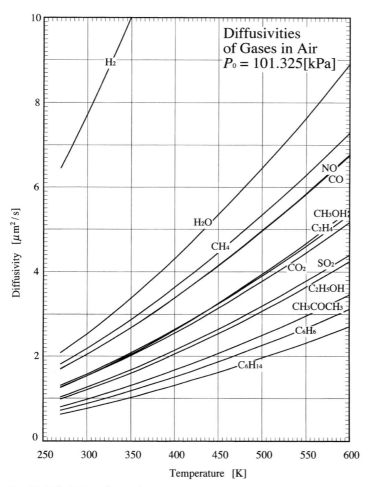

Fig. A5 Diffusivities of gases in air.

Subject Index

a

accumulation
- of non-condensable gas   192 f., 203
activity coefficient   7
air separation plant   249
Antoine's equation   6, 134, 214, 233
apparent end effect   227 f.
argon column   249, 251
aspect ratio   116

b

Bassel function   93
binary diffusion   12
- coefficient   12, 14 f.
- flux   12 ff.
binary distillation physical picture   216
- effect of partial condensation   218 ff.
- vapor phase temperature distribution   218
Blasius' empirical correlation   154
Blasius' equation   154
blowing parameters   82, 172
boundary layer   53
- concentration distribution   64
- polynominal concentration approximation   71
- polynominal velocity approximation   70
- velocity distribution   57
boundary layer equation   53
- dimensionless   56
- integral form   57
- numerical solution   62
bubble
- continuous phase mass transfer   140
- point   189, 217, 234
- shape   121 f.
Buckingham's Pi theorem   50
buffer layer   152

c

calculation of ternary distillations   239 ff.
Chilton and Colburn's analogy   113, 158 ff.
- comparison with data   159
circulation flow, effect of   140
circulation velocity   124
classical model for distillation   212, 214
concentration
- binary system   22
- multi-component system   23
concentration distribution
- boundary layer, integral form   61
- one-seventh law   171
concentration driving force   20, 228
- binary distillation   220
condensation heat transfer coefficient   186
- circular cylinder   201
- total condensation   191 f.
condensation
- dropwise   183
- of a vapor   192
condensation with non-condensable gas
- heat and mass transfer   193
- horizontal tube   203
conduction of heat   4
continuity equation   28 f.
continuous phase   122
continuous phasestream function   122
convective mass flux   3, 17, 20
- binary distillation   219
- gas absorption   221
- multi-component distillation   221
- ternary distillation   220
convective velocity   18
- correlation, cryogenic flux   249
creeping flow   101 ff., 122
- stream function   102
cryogenic distillation   249 ff.
- cylindrical coordinate   260

Mass Transfer. From Fundamentals to Modern Industrial Applications. Kenichi Asano
Copyright © 2006 WILEY-VCH Verlag GmbH & Co. KGaA, Weinheim
ISBN: 3-527-31460-1

d

Deissler analogy 153, 162 ff.
dew point 217, 234
– curve 189
diffusion 5, 9, 21
– coefficient 265
– – multi-component 15
diffusion equation 29 ff.
– boundary layer 55, 60
– male fraction 31
– mass fraction 30
diffusional flux 3, 9, 14, 17
– correlation of binary distillation 228
– correlation of ternary distillation 229
– binary distillation 219
– mass 11, 13
– molar 11 f.
– ternary distillation 22
– turbulent 148
diffusivity 6
– turbulent 149
dimensional analysis 47 ff.
dimensional homogeneity 47 f.
dimensionless
– concentration 59
– number 44 ff.
– dimensionless 59
dispersed phase 122
– mass transfer 141
– Sherwood number 137
– stream function 122
displacement thickness 168
distillation
– binary, non-adiabatic 221
– classical model 212, 214
– path 213, 232
– rate controlling process 217
– setup, double-column 249
– ternary, non-adiabatic 239
distillation calculation, multi-component
– convergence of terminal composition 242
– flow chart 242
drag coefficient 103
– bubble 125
– correlation 105
– drop in gas 125
– effect of mass injection or suction 126
– Hadamard's flow 123
– interaction effect 117
– numerical solution 104 f.
– spheroidal particle 115
dropwise condensation 183
dumping factor 150, 153

dynamic simulator, oxygen 18
– separation process 255

e

eddy
– diffusional flux 148
– diffusivity 149
– heat flux 147
– kinematic viscosity 148
– thermal diffusivity 149
effective diffusion coefficient 15, 235
effective interfacial area 225/226
energy equation 33
– boundary layer 55, 60
enriching section 241
equation of motion 33
equilibrium, local 20, 178
equilibrium stage 241
– model 209
equimolal counterdiffusion 18 f.
error function 35
Eucken's equation 265
evaporating drop 126
– drag coefficient 131
– falling freely in gas 132 ff.
– heat and mass transfer 128, 131
– simulation 135
evaporation of fuel spray 126

f

falling liquid film 73
– gas absorption, long exposure time 76 ff.
– thickness 75
– velocity distribution 73 ff.
Fick's Law 5, 10, 12 f., 30 f.
– film condensation 183
– physical representation 184
– pure vapor, circular cylinder 200 f.
– variable physical property 187
film model 34 f.
flow around and evaporating drop 126 ff.
fluctuation velocity 145
fluid friction 3
fluidized bed
– heat and mass transfer 117
form drag 103
forward stagnation point 112, 118
Fourier number 137
Fourier's law 4
free stream 53
free surfaces 121
friction factor 58 f.
– average
– – laminary boundary layer 59

– – turbulent boundary layer   170
– circular pipe   90
– local
– – laminar boundary layer   59
– – turbulent boundary layer   169 f.
– turbulent flow, circular pipe   153 ff.
friction velocity   151
frictional drag   102
fundamental dimension   47

g
gas absorption
– falling liquid film, short contact time   75
governing equation
– cylindrical coordinate   260
– spherical coordinate   261
Graetz number   223
– for heat transfer   91
– for mass transfer   92, 99
Graetz's problem   95

h
Hadamard's flow   122
Hausen's approximation   96
heat conduction equation   93
heat flux
– turbulent   147
heat transfer boundary layer   65
– small Prandtl number   69
heat transfer entrance region
– circular cylinder   201
– circular pipe   93 f.
– fully developed flow
– – circular pipe   95
– interaction effect, two particle   117
– spheroidal particle   115
– – stationary fluid   109 ff.
– turbulent boundary layer   171
– – high mass flux effect   202
Henry's constant   6, 262
Henry's law   6
Higbie's model (see penetration model)
Higbie's penetration model   75
high mass flux effect   3, 80, 126, 128, 130, 172, 220
– correlation   84
– drag coefficient   131
– friction factor   172
– heat and mass transfer of drop   132
– heat transfer   172
– mass transfer, turbulent boundary effect   173
– numerical solution   84

high-pressure column   249
Hirshfelder's equation   235
– viscosity   265
– diffusion coefficient   265
HTU   209, 211, 214, 221
humidity chart   264

i
ideal solution   6
incompressible fluid   29
interfacial velocity   18, 20
isotope-exchange reactor   254

j
j-factors   158

k
Kronig–Brink model   141

l
laminar flow   27, 90
laminar sub-leyer   152
Lapple–Shepard correlation
– drag coefficient   106
latent heat   177
law of conservation
– of energy   33
– of mass   28 f.
Leveque's approximation   96
Lewis number   180
local distribution of heat
– diffusional flux, spherical particle   112
logarithmic mean concentration driving force   92
logarithmic mean temperature driving force   92
logarithmic velocity distribution layer   168
low-pressure column   249, 251

m
mass average velocity   11, 13
mass flux   16
– cryogenic distillation   249
mass transfer   16
– boundary layer   65
– circular cylinder   201
– – high mass flux effect   202
– coefficient   20, 43
– – overall   24
– definition   23
– entrance region, circular pipe   93 f.
– fully developed flow, circular pipe   95
– inactive entrance region   72
– interaction effect, two particle   117

- large Schmid number 67
- spherical particle, stationary fluid 109 ff.
- spheroidal particle 115
- tray tower 210
- turbulent boundary layer 171
- wetted-wall column 97 ff.
θ-method of convergence 245
Mickley's film model 81 ff., 129, 172
mixing length 149 f.
molar average velocity 13, 20
molar flux 17
momentum thickness 168
motion of drop
- gas phase 125

n
Navier–Stokes equation 33
Newton's law of viscosity 3 f.
Newtonian fluids 4
non-ideal solution 6
non-Newtonian fluids 4
number of transfer unit 211, 214
- discontinuity 213
Nusselt equation, falling liquid film 75
Nusselt number 41
- local 60
Nusselt's model 183 ff., 195

o
oblate spheroid 121
one-nth power law 151, 168
one-seventh power law 151
operating line 214
Othmer's experiment 194 ff.
oxygen-18 252
- separation process 253 ff.
- – flow diagram 254

p
packed columns 209
- calculation, finite reflux ratio 244
parabolic velocity distribution
- circular pipe 89
partial condensation 189, 220
- binary distillation 229
particle Reynolds number 103
penetration model 35 f.
PET diagnostics 253
phase equilibrium 2
phase rule 178
Prandtl number 39 ff.
psychrometric ratio 180
- correlation 181

q
quasi-linearization method 62

r
Ranz–Marshall correlation 112 ff.
Raoult's law 6, 233
Rayleigh's method of indices 49 f.
reboiler 241
reboiler-condenser 249
reference temperature 187
reference velocity diffusion 11
Reynolds analogy 157 f.
Reynolds number
- average, boundary layer 58
- critical 39, 145
- local, boundary layer 58
- particle 103
- physical interpretation 38 f.
Reynolds stress 146 ff.

s
saturated temperature 177
Schmidt number 39 ff.
- multi-component 230
Sherwood number 42 f.
- local 61
- mass 42
- molar 42
- mutual relation 43
similarity transformation
- boundary layer equation 55
similitude, principle 47 f.
simulation
- pilot-plant scale air separation 251
simultaneous heat
- mass transfer model 177, 179, 209, 231 f., 251, 254
solid wall dissolution
- falling liquid film 78 ff.
solid-sphere penetration model 136
spheroidal particle 115
spray absorption 136
Stokes' flow 103 f., 115
- liquid phase mass transfer 115
Stokes' law of resistance 103
stream function 56
- dimensionless 56
stripping section 241
summation theorem
- for mass diffusional flux 17
super cryogenic distillation 255
surface-contaminated fluid spheres
- motion 126
surface renewal model 36 f.

surface temperature
– evaporation   178 f.

t

temperature boundary layer   39 f.
temperature distribution
– boundary layer, integral form   60
– near the interface   177
terminal velocity   106 ff., 124, 135
– bubble   125
– calculation   107
– drop in gas   125
ternary distillations
– calculation   239 ff.
ternary packed column distillation
– total flux calculation   233
thermal conductivity   5
thermal diffusivity   148
– turbulent   149
time-averaged velocity   145
total condensation   189
– vapor phase temperature distribution   190
transfer number   83, 128
– for heat transfer   85
– for mass transfer   85
tray efficiency   209, 221
– discontinuity   212
– Murphree   211
tray tower   209
turbulent boundary layer
– velocity distribution   168

turbulent core   152
turbulent flow   27, 90, 145

u

ultra-pure grade oxygen   254
unidirectional diffusion   18, 34
universal velocity distribution law   153
– circular pipe   151 ff.

v

vapor-liquid equlibrium   214
velocity boundary layer   38
viscosity   3
void function   117, 119
– simulation   118
voidage   119
volume average diameter   117
von Ka'rman analogy   161

w

wake   118
wet-bulb temperature   134, 179
– evaporating drop   135
wet-bulb temperature   179
wetted area   225
– metal-structured packing   226
wetted-wall column, binary distillation   222
Wilke's effective diffusion coefficient   230, 251
Wilke's equation   265
Wilson's equation   214